Transmission Electron Microscopy

III Imaging

Library of Congress Cataloging in Publication Data

Williams, David B. (David Bernard), 1949–
  Transmission electron microscopy: a textbook for materials science / David B. Williams and C. Barry Carter.
     p.   cm.
  Includes bibliographical references and index.
  ISBN 0-306-45247-2 (hardbound).—ISBN 0-306-45324-X (pbk.)
  1. Materials—Microscopy. 2. Transmission electron microscopy. I. Carter, C. Barry. II. Title.
TA417.23.W56 1996   96-28435
502′.8′25—dc20   CIP

ISBN 0-306-45247-2 (Hardbound)
ISBN 0-306-45324-X (Paperback)

© 1996 Plenum Press, New York
A Division of Plenum Publishing Corporation
233 Spring Street, New York, N. Y. 10013

10 9 8 7 6 5 4 3 2 ·

All rights reserved

No part of this book may be reproduced, stored in a retrieval system, or transmitted in any form or by any means, electronic, mechanical, photocopying, microfilming, recording, or otherwise, without written permission from the Publisher

Printed in the United States of America

Imaging in the TEM

22

- 22.1. What Is Contrast? 351
- 22.2. Principles of Image Contrast 351
  - 22.2.A. Images and Diffraction Patterns 351
  - 22.2.B. Use of the Objective Aperture or the STEM Detector: BF and DF Images 351
- 22.3. Mass-Thickness Contrast 353
  - 22.3.A. Mechanism of Mass-Thickness Contrast 353
  - 22.3.B. TEM Images 354
  - 22.3.C. STEM Images 355
  - 22.3.D. Specimens Which Show Mass-Thickness Contrast 357
  - 22.3.E. Quantitative Mass-Thickness Contrast 357
- 22.4. Z Contrast 358
- 22.5. TEM Diffraction Contrast 361
  - 22.5.A. Two-Beam Conditions 361
  - 22.5.B. Setting the Deviation Parameter, s 361
  - 22.5.C. Setting Up a Two-Beam CDF Image 361
  - 22.5.D. Relationship between the Image and the Diffraction Pattern 363
- 22.6. STEM Diffraction Contrast 364

CHAPTER PREVIEW

We've already mentioned back in Chapters 2–4 that TEM image contrast arises because of the scattering of the incident beam by the specimen. The electron wave can change both its amplitude and its phase as it traverses the specimen and both these kinds of change can give rise to image contrast. Thus a fundamental distinction we make in the TEM is between *amplitude contrast* and *phase contrast*. In most situations, both types of contrast actually contribute to an image, although one will tend to dominate. In this chapter we'll discuss only amplitude contrast and we'll see that there are two principal types, namely *mass-thickness contrast* and *diffraction contrast*. This kind of contrast is observed in both TEM and STEM BF and DF images and we'll discuss the important differences between the images formed in each of these two modes of operation. We'll then go on to discuss the principles of diffraction contrast, which are sufficiently complex that it takes Chapters 23–26 to show you how this form of contrast is used to identify and distinguish different crystal defects. Diffraction-contrast imaging came into prominence in about 1956, when it was realized that the intensity in a diffracted beam depends strongly on the deviation parameter, **s**, and that crystal defects distort the diffracting planes. Therefore, the diffraction contrast from regions close to the defect would depend on the properties (in particular, the strain field) of the defect. We'll then consider phase contrast and how it can be used to image atomic level detail in Chapters 27–30. Other forms of TEM imaging and variations on these major types of contrast are gathered in the catch-all Chapter 31.

Imaging in the TEM 22

22.1. WHAT IS CONTRAST?

Before we start to describe specific types of contrast it's worth a quick reminder of what exactly we mean by the word "contrast." We can define contrast (C) quantitatively in terms of the *difference* in intensity (ΔI) between two adjacent areas

$$C = \frac{(I_1 - I_2)}{I_2} = \frac{\Delta I}{I_2} \qquad [22.1]$$

In practice your eyes can't detect intensity changes < 5%, and even < 10% is difficult. So unless the contrast from your specimen exceeds > 5–10% you won't see anything on the screen or on the photograph. However, if your image is digitally recorded, you can enhance low contrast electronically to levels at which your eyes can perceive it. We'll return to image processing and contrast enhancement in Chapter 30.

So we see contrast in TEM images as different levels of green light coming from the viewing screen or CRT. On the photograph, contrast is seen as different gray levels and our eyes can only discern about 16 of these. If we want to quantify the contrast, we need to make direct intensity measurements, e.g., via a microdensitometer, but usually it's only necessary to see qualitative differences in intensity. Be careful not to confuse *intensity* with *contrast* when you describe your images. We can have strong or weak contrast but not bright or dark contrast. The terms "bright" and "dark" refer to density (number/unit area) of electrons hitting the screen/detector, and the subsequent light emission that we see. In fact, you generally get the strongest contrast under illumination conditions that lower the overall intensity, while if you try and increase the number of electrons falling on the screen, by condensing the beam onto a reduced area of the specimen, you'll usually lower the image contrast. These points are summarized in Figure 22.1, which defines intensity and contrast.

Before we discuss the two forms of amplitude contrast in detail, we need to remind you of the operational principles for creating amplitude contrast in your image. We obtain contrast in our images either by selecting specific electrons or excluding them from the imaging system. We have two choices: we can form either BF or DF images by selecting the direct or scattered electrons, respectively. So this chapter builds on what you learned about electron scattering in Chapters 2–4 and how to operate the TEM, described in Chapter 9.

22.2. PRINCIPLES OF IMAGE CONTRAST

22.2.A. Images and Diffraction Patterns

If you go back and look at Figure 2.1, you'll see that the uniform electron intensity in the incident beam is transformed into a nonuniform intensity after scattering by the specimen. So a variable electron intensity hits the viewing screen or the electron detector, which translates into contrast on the screen or CRT. Now you also know that the DP shows you this nonuniformity because it separates out the diffracted and direct beams. Therefore, a fundamental principle of imaging in the TEM is: first *view the DP*, since this pattern tells you how your specimen is scattering. The relationship between the image and the DP is most critical for crystalline specimens showing diffraction contrast. However, you need to view the DP first, whatever contrast mechanism you want to use, and whatever specimen you are studying.

22.2.B. Use of the Objective Aperture or the STEM Detector: BF and DF Images

In order to translate the electron scatter into interpretable amplitude contrast we select *either* the direct beam or some of the diffracted beams in the SAD pattern to form BF and

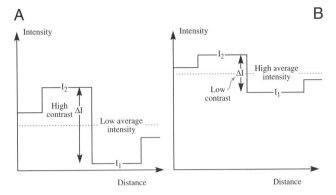

Figure 22.1. Schematic intensity profiles across an image showing (A) different intensity levels (I_1 and I_2) and the difference (ΔI) between them, which defines the contrast. Generally, in a TEM, if the overall intensity is increased (B) the contrast decreases.

DF images, respectively. Note we are justified in using the "beam" terminology, since the electrons have left the specimen. We've already seen back in Section 9.3 that in a TEM we select the direct or a scattered electron beam with the objective aperture. Remember, if you form an image without the aperture, the contrast will be poor because many beams then contribute to the image. Furthermore, aberrations due to the off-axis electrons will make your image impossible to focus. Your choice of the aperture size governs which electrons contribute to the image and thus you control the contrast.

Figure 22.2 shows a DP from a single-crystal Al specimen with the schematic indication of the objective aperture indicated. In this figure, the aperture in position A is selecting the direct beam only and thus a BF image will be formed in the image plane of the lens. This arrangement will produce amplitude contrast whether the specimen is crystalline (as in this case) or amorphous. If the aperture is in position B, it will select only electrons scattered in that specific direction. Thus a DF image will be formed. Usually

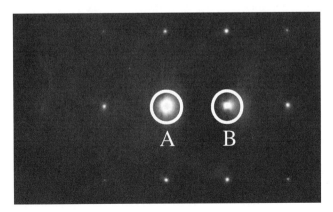

Figure 22.2. The relationship between the objective aperture and the diffraction pattern for forming (A) BF and (B) DF images.

we tilt the incident beam such that the scattered electrons remain on axis, creating a centered dark-field (CDF) image, which we described back in Section 9.3. We'll discuss CDF techniques later in Section 22.5 and we'll usually assume CDF is the operational mode in DF imaging. *Thus, BF and DF are the two basic ways to form amplitude-contrast images.* We'll see later that if you want to observe phase contrast, you have to use an objective aperture that is large enough to gather more than one beam.

In a STEM we select the direct or scattered beams in an equivalent way but use detectors rather than apertures. We compare the two different operational modes in Figure 22.3. Again, we saw back in Section 9.4 that we insert a BF on-axis detector, or an annular DF (ADF) detector, in a conjugate plane to the back focal plane. We control which electrons fall on which detector and thus contribute to the image by adjusting the post-specimen (imaging)

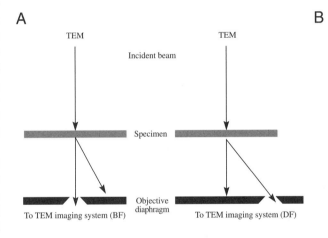

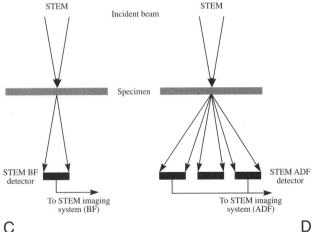

Figure 22.3. Comparison of the use of an objective aperture in TEM to select (A) the direct or (B) the scattered electrons forming BF and DF images, respectively. In STEM we use (C) an on-axis detector or (D) an annular detector to perform equivalent operations.

lenses to change the camera length. Clearly, for DF imaging, the ADF detector gathers many more electrons than the objective aperture, which is good for imaging some specimens and bad for imaging others, as we'll see.

So, in summary, we can create BF or DF images with the direct beam or scattered beams, respectively. In order to understand and control the contrast in these images you need to know what features of a specimen cause scattering and what aspects of TEM operation affect the contrast.

22.3. MASS-THICKNESS CONTRAST

Mass-thickness contrast arises from incoherent (Rutherford) elastic scatter of electrons. As we saw back in Chapter 3, the cross section for Rutherford scatter is a strong function of the atomic number Z, i.e., the mass or the density, ρ, as well as the thickness, t, of the specimen. Rutherford scattering in thin specimens is strongly forward peaked. Therefore, if we form an image with electrons scattered at low angles (<~5°), mass-thickness contrast dominates (but it also competes with Bragg-diffraction contrast). However, we'll also see that at high angles (>5°), where coherent scattering is negligible, we can pick up low-intensity, incoherently scattered beams. The intensity of these beams depends on atomic number (Z) only. Thus we can also get so-called Z contrast, which contains elemental information like that in BSE images in the SEM; in a DSTEM, however, we can obtain these images with atomic resolution. It is also feasible to form BSE images in a TEM but, because the specimen is thin, the number of BSEs is so small that the images are noisy and of poor quality, so no one does it. You shouldn't waste your money buying a BSE detector!

Mass-thickness contrast is most important if you are looking at noncrystalline materials such as polymers and it is *the* critical contrast mechanism for biological scientists. But as we'll see, any variations in mass and thickness will cause contrast. As you learned in Chapter 10, it's almost impossible to thin a bulk sample uniformly and so all real specimens will show some mass-thickness contrast. In some cases this will be the only contrast you can see.

In this section, we'll assume that there is no contribution to the image from diffraction contrast. This is automatically so if the specimen is amorphous. If the specimen is crystalline, then remove the objective aperture or use the ADF detector to gather many beams and thus minimize any diffraction contrast. As you'll see, you should still use an objective aperture to enhance the mass-thickness contrast, i.e., you'll still create BF and DF images of amorphous materials.

22.3.A. Mechanism of Mass-Thickness Contrast

The mechanism by which differences in mass and thickness cause contrast is shown in Figure 22.4 and at this stage we'll talk about the process qualitatively. As electrons go through the specimen they are scattered off axis by elastic nuclear interactions, i.e., Rutherford scattering. You know two factors from Chapter 3:

- The cross section for elastic scattering is a function of Z.
- As the thickness of the specimen increases, there will be more elastic scattering because the mean-free path remains fixed.

So using a very simple, qualitative argument you would expect high-Z (i.e., high-mass) regions of a specimen to scatter more electrons than low-Z regions of the same thickness. Similarly, thicker regions will scatter more electrons than thinner regions of the same average Z, all other factors being constant. Usually, mass-thickness contrast images are interpreted in such a purely qualitative fashion, although we'll see a little later that it is possible to quantify the scattering intensity. So, as you can see from Figure 22.4, for the case of a BF image, thicker and/or higher-mass areas will appear darker than thinner and/or lower-mass areas. The reverse will be true for a DF image.

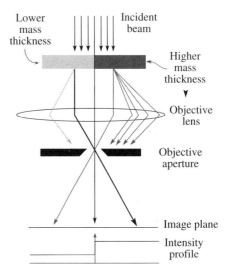

Figure 22.4. Mechanism of mass-thickness contrast in a BF image. Thicker or higher-Z areas of the specimen (darker) will scatter more electrons off axis than thinner or lower-mass (lighter) areas. Thus fewer electrons from the darker region fall on the equivalent area of the image plane (and subsequently the screen), which therefore appears darker in BF images.

This is all you need to know to interpret mass-thickness contrast images. Sometimes mass-thickness contrast is explained in terms of different amounts of electron absorption within the specimen and so you may come across the expression "absorption contrast." We think that this term is misleading, because in thin foils the actual amount of electron absorption is small; scattering outside the aperture or the detector, not absorption within the specimen, causes the contrast. For much the same reason, we prefer not to use the term "structure-factor contrast," which is sometimes used to describe this phenomenon, since this implies a Bragg contribution, which may or may not be present.

However, you should be aware that if there are small crystals of different atoms in a given foil thickness, differences in their structure factor (F) from that of the matrix will cause contrast changes, since $I \propto |F|^2$. For example, you can detect the presence of nanometer-size clusters of Ag atoms in very thin foils of Al alloys in this way. Conversely, an absence of atoms (e.g., a void) will also scatter differently, although Fresnel contrast (see Chapter 27) is a better way to detect voids and bubbles.

Let's first look at a few images showing mass-thickness contrast and see which TEM variables you can control.

22.3.B. TEM Images

Figure 22.5A is a TEM BF image of some latex particles on an amorphous-carbon support film. Assuming the latex is predominantly carbon, we have a constant Z and varying t. So the latex particles are darker than the support film since they are thicker. What you are basically seeing is a shadow projection image of the latex particles. Because it is a projection image, you cannot say that the particles are spheres (which in fact they are). They could equally well be disks or cylinders. To tell something about the shape, you need to shadow them, i.e., evaporate a thin heavy metal (Au or Au-Pd) coating at an oblique angle as shown in Figure 22.5B. The shadow shape reveals the true shape of the particles (Watt 1985).

Shadowing introduces some mass contrast to what was just a thickness-contrast image. If we assume the Au-Pd film is very thin compared to the carbon support film, then the contrast across the edge of the shadow is predominantly mass contrast, due to the difference in average Z of the Au-Pd and the carbon film. There is also an intensity change across the latex spheres reflecting the preferential deposit of Au-Pd on the side of the sphere toward the source of the evaporated metal.

It is an intriguing exercise to print Figure 22.5B in reverse (or take a DF image), as shown in Figure 22.5C. In this image, the latex spheres now appear to stand proud

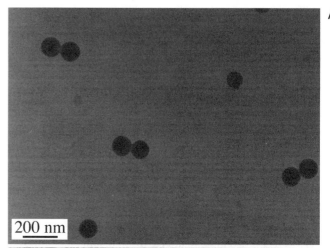
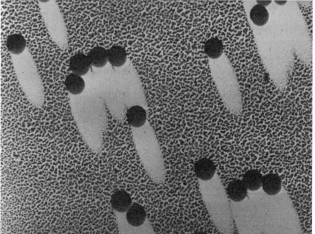
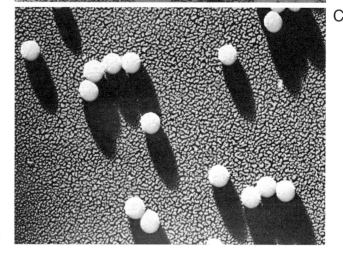

Figure 22.5. (A) TEM BF image of latex particles on a carbon support film showing thickness contrast only. (B) Latex particles on a carbon film shadowed to reveal the shape of the particles through the addition of selective mass contrast to the image. (C) Reverse print of (B) exhibits a 3D appearance.

of the surface, even though you're still viewing a two-dimensional projected image. Because the shadows are now dark, your brain interprets the picture as it would a reflected-light image and endows it with a 3D nature. While the interpretation in this case is correct, it may not always be so. Once again we stress that you must be careful when interpreting two-dimensional images of 3D specimens.

In addition to the use of shadowing to enhance mass-thickness contrast, it is common practice to stain different areas of polymers and biological specimens with heavy metals such as Os, Pb, and U (Sawyer and Grubb 1987). The stain leaves the heavy metal in specific regions of the structure (e.g., at unsaturated C=C bonds in a polymer and cell walls in biological tissue) and therefore these areas appear darker in a BF image. Figure 22.6 shows a BF image of a stained two-phase polymer. Since the specimen is of constant thickness (it was ultramicrotomed) the image shows mass contrast only.

> The TEM variables which affect the mass-thickness contrast for a given specimen are the objective aperture size and the kV.

If you select a larger aperture, you allow more scattered electrons to contribute to the BF image. So the contrast between scattering and nonscattering areas is lowered, although the overall image intensity increases. If you choose a lower kV, both the scattering angle and the cross section increase. Hence more electrons will be scattered outside a given aperture, hitting the diaphragm, and contrast will increase at the expense of intensity. The decrease in intensity will be worse for TEMs with a thermionic source because the gun brightness decreases as the kV is lowered. Figure 22.7 shows how a smaller aperture size results in improved contrast. Of course, any decrease in intensity can be offset by increased exposure times until specimen drift becomes a limiting factor.

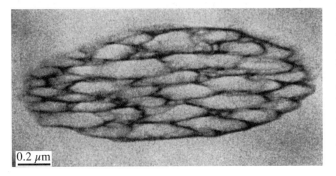

Figure 22.6. BF image of stained two-phase polymer exhibiting mass contrast due to the segregation of the heavy metal atoms to the unsaturated bonds in the darker phase.

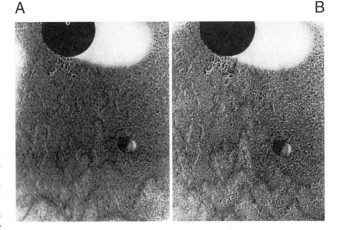

Figure 22.7. The effect of objective aperture size on mass-thickness contrast; the images of the shadowed latex particle were taken with an aperture size of (A) 70 μm and (B) 10 μm. A smaller aperture enhances the contrast, in a similar manner to lowering the kV.

Now for DF images, there isn't much more to be said and the images will generally show complementary contrast to BF, as we saw in Figures 22.5B and C. The overall intensity of the DF image will be much lower than the BF image (hence the relative terms "dark" and "bright") because the objective aperture will select only a small fraction of the scattered electrons. It's easy to remember that the BF image of a hole in your specimen will be bright and a DF image will be dark. However, remember that the corollary of low intensity is high contrast, and DF images generally show excellent contrast.

22.3.C. STEM Images

In a STEM you have more flexibility than in a TEM because, by varying L, you change the collection angle of your detector and create, in effect, a variable objective aperture. So you have more control over which electrons contribute to the image. Even so, STEM BF images offer little more than TEM BF images. Generally, STEM images are noisier than TEM images (unless you've got an FEG STEM). Figure 22.8 shows a noisy STEM BF image of the same two-phase polymer as shown in the TEM image in Figure 22.6. The STEM image generally shows poorer resolution because, with good thin specimens, the beam size dominates the resolution. To get reasonable intensity in a scanning image in reasonable time we have to use a large beam, as we discussed when we compared scanning and static images back in Chapter 9. Figure 22.9 shows the difference between (A) TEM and (B) STEM BF images from a low-contrast specimen. The STEM image contrast has been enhanced and is considerably greater than in the TEM image, but the noise in the image is also more visible. However, if

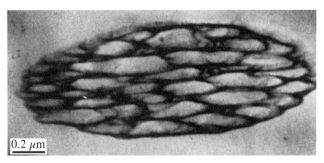

Figure 22.8. STEM BF image of a stained two-phase polymer. Comparison with the TEM image in Figure 22.6 shows that, while the contrast is higher in STEM, the image resolution is poorer.

you record your TEM image using a CCD camera, or digitize the negative, you can enhance the contrast (see Chapter 30). A good way to do this is with differential hysteresis imaging (Section 1.5), as you can see in Figure 22.9C.

> Remember, you can always increase the contrast in the STEM image by adjusting the signal-processing controls, such as the detector gain and black level and the contrast and brightness controls on the CRT; such options aren't available for analog TEM images.

In a STEM the scattered electrons fall onto the ADF detector. This gives rise to a fundamental difference between the TEM and STEM DF modes:

- DF TEM images are usually formed by permitting only a fraction of the scattered electrons to enter the objective aperture.
- STEM images are formed by collecting most of the scattered electrons on the ADF.

Therefore, STEM ADF images are less noisy than TEM DF images, as shown in Figure 22.10. Because lenses aren't used to form the STEM image, the ADF images don't suffer aberrations, as would the equivalent off-axis TEM DF image.

STEM ADF image contrast is greater than TEM DF contrast: in STEM, L can be adjusted to maximize the ratio of the number of scattered electrons hitting the detector to the number of electrons going through the hole in the middle of the detector. You can thus improve the contrast quite easily, just by watching the CRT and adjusting L.

> The STEM must be well aligned so the DP expands and contracts on axis.

However, as you can see from Figure 22.10, while the TEM DF image shows poorer contrast and is noisier, it still shows better resolution. STEM images generally only show better resolution than TEM images when thick specimens are being imaged, because the chromatic aberration effects from thicker specimens do not affect the STEM images. If contrast is more important than resolution, then STEM is more useful. Indeed, in a STEM, you can study unstained polymer specimens which would show negligible contrast in a TEM.

STEM imaging is also useful if your specimen is beam sensitive, e.g., some polymers. A scanning beam lets you precisely control the irradiated region of the specimen, so it's a form of low-dose microscopy (see Section 4.6). You'll lose some image resolution unless you have access to an FEG STEM.

The comparison we've made of TEM and STEM images here is qualitative, but there have been many rigorous comparisons of STEM and TEM contrast, particularly

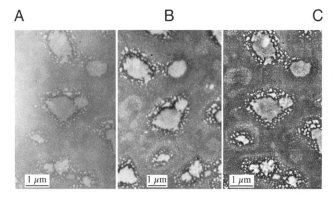

Figure 22.9. Comparison of TEM (A) and STEM (B) images of an amorphous SiO_2 specimen containing Cl-rich bubbles. The low mass contrast in the TEM image can be enhanced in a STEM image through signal processing. (C) A similar effect can be achieved by digitizing the TEM image (A) and applying contrast-enhancement software.

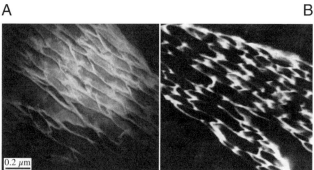

Figure 22.10. Comparison of (A) TEM DF and (B) STEM ADF images of the same two-phase polymer as in Figures 22.6 and 22.8. As in BF the STEM image shows higher contrast but lower resolution. Also, the ADF aperture collects more signal than the TEM objective aperture so the STEM image is less noisy.

for biological specimens (e.g., Cosslett 1979). When STEMs were first introduced in the 1970s, the absence of chromatic aberration effects led to prophesies that STEM image resolution would invariably be better than TEM; there were even predictions of the end of classical TEM imaging! This hasn't happened because, as we'll see, there is more than just the chromatic aberration factor that governs the image quality, particularly for crystalline specimens. In summary, then, there are three reasons to use STEM mass-thickness contrast images:

- The specimen is so thick that chromatic aberration limits the TEM resolution.
- The specimen is beam-sensitive.
- The specimen has inherently low contrast in TEM, and you can't digitize your TEM image or negative.

22.3.D. Specimens Which Show Mass-Thickness Contrast

Mass-thickness is the primary contrast source in amorphous materials, which is why we've illustrated this portion of the chapter mainly with polymer specimens. Replicas also display thickness contrast (see Figure 22.11A). Remember from Chapter 10 that replicas recreate the specimen topography, e.g., for a fracture surface. The amorphous-carbon replica can be unshadowed (Figure 22.11A) or shadowed (Figure 22.11B). The uneven metal shadowing increases the mass contrast and thus accentuates the topography; see also the latex particles in Figure 22.3. An extraction replica (Figure 22.11C) or particles dispersed on a support film will also show mass-thickness contrast, and shadowing could be useful to reveal the shape of the particles. If the particles are crystalline there will also be a component of diffraction contrast.

22.3.E. Quantitative Mass-Thickness Contrast

Because mass-thickness contrast is governed by Rutherford scattering, we can use the equations given back in Chapter 3 to predict the effect of Z and t on the scattering angle, θ, and the effect of kV on the cross section. We assume that the atoms scatter independently (i.e., the scattering is truly incoherent). This is not the case, since even DPs from amorphous specimens show diffuse rings rather than uniform intensity (Figure 2.5D). Nevertheless, we'll still assume incoherent scattering.

As we stated at the start of this chapter, the contrast C is given by $\Delta I/I$ and it can be shown (e.g., Heidenreich

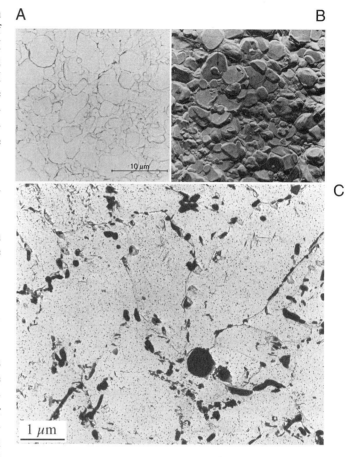

Figure 22.11. More examples of mass-thickness contrast: (A) a carbon replica of a fracture surface doesn't show much of either form of contrast until (B) oblique shadowing enhances the topography. (C) An extraction replica of a range of small precipitate particles in a Cr-Mo steel weld shows both mass and thickness contrast.

1964) that a change in thickness, Δt, at constant atomic number Z creates contrast

$$\frac{\Delta I}{I} = 1 - e^{-Q\Delta t} = Q\Delta t \qquad [22.2]$$

for $Q\Delta t < 1$, where Q is the total elastic scattering cross section. Since the minimum contrast we can see is ~5%, then the minimum Δt that we can see is

$$\Delta t \cong \frac{5}{100\, Q} = \frac{5\, A}{100\, N_0 \sigma \rho} \qquad [22.3]$$

where A is the atomic weight, N_0 is Avogadro's number, σ is the single-atom scattering cross section, and ρ is the density.

A similar argument can be made if there is a ΔZ (in which case σ changes). So, if we want to calculate the contrast, we need to know σ. As we've seen in equation 3.9, for low-angle scattering, the differential Rutherford cross section is equal to $f(\theta)^2$ where $f(\theta)$ is the atomic scattering factor, given by equation 3.10

$$f(\theta) = \frac{\left(1 + \frac{E}{E_0}\right)}{8\pi^2 a_0} \left(\frac{\lambda}{\sin\frac{\theta}{2}}\right)^2 (Z - f_x) \quad [22.4]$$

The Z term represents the Rutherford scattering. For unscreened Rutherford scattering (i.e., ignoring the effects of the electron cloud) σ is proportional to Z^2. This unscreened behavior is approximated by electrons scattered through semiangles above ~5° (for Cu) although it is dependent on E_0 and Z. At lower angles, scattering becomes increasingly screened, less dependent on Z, and more dominated by inelastic scattering and diffraction. There is no precise angle which we use to define the transition from low- to high-angle scatter, but the effect of screening effectively disappears at angles > θ_0, the screening parameter, defined back in equation 3.6.

We can use the unscreened Rutherford scattering expression to determine the probability that an electron will be scattered through greater than a given angle. To do this, we integrate the differential Rutherford cross section from an angle β (defined by the semiangle of collection of the objective aperture) to infinity. Thus

$$\sigma(\beta) = 2\pi \int_\beta^\infty |f(\theta)|^2 \theta\, d\theta \quad [22.5]$$

which can be evaluated (see Reimer 1993) to give

$$\sigma(\beta) = \frac{\left[Z\lambda\left(\frac{a_0}{Z^{0.33}}\right)\left(1 + \frac{E}{E_0}\right)\right]^2}{\pi(a_0)^2 \left(1 + \left(\frac{\beta}{\theta_0}\right)^2\right)} \quad [22.6]$$

where a_0 is the Bohr radius and θ_0 is the characteristic screening angle; all the other terms have their usual meaning (see Chapter 3). So in equation 22.6 you can see directly the effect of Z and kV on electron scatter and hence on contrast. As we've already described, higher-Z specimens scatter more while lowering E_0 increases scattering. The effect of thickness is deduced from the mean-free path for elastic scatter, λ (which is inversely proportional to σ). So, thicker specimens scatter more.

Let's assume that n electrons are incident on the specimen and dn electrons are scattered through an angle > β. Then, from equation 22.6, ignoring any inelastic scattering (which isn't really reasonable, but we'll do it to simplify matters), the reduction in the number of electrons going through the objective aperture to form the BF image is given by

$$\frac{dn}{n} = -N\sigma(\beta)\, dx \quad [22.7]$$

where $N = N_0/A$ and N_0 is Avogadro's number; $\sigma(\beta)$ is given by equation 22.6 and $x = \rho t$. So this expression gives the dependence of the contrast on Z and t. If we integrate

$$\ln n = N\sigma x + \ln n_0 \quad [22.8]$$

and if we rearrange this expression, we obtain

$$n = n_0 e^{-N\sigma x} \quad [22.9]$$

which describes the exponential decrease in the number of scattered electrons as the specimen mass-thickness ($x = \rho t$) increases.

As you'll have gathered, this equation is somewhat of an approximation but it does give you a feel for the factors that control mass-thickness contrast. For a given specimen, the variables are local changes in Z and t and within the microscope the variables are β and E_0, which you can control to change the contrast as we saw in Figure 22.7.

In principle, you could use these equations and equation 22.1 to calculate the expected contrast arising from differences in Z or t and see if they were detectable at the 5% contrast level. In practice, however, image contrast calculations are not carried out for simple mass-thickness contrast in materials specimens.

22.4. Z CONTRAST

Z contrast is the name given to a high-resolution (atomic) imaging technique, but rather than discuss it in Chapter 28 under phase-contrast effects we'll talk about it here, because it represents the limit of mass-thickness contrast where detectable scattering arises from single atoms or a column of atoms.

Back in the 1970s, early FEG STEMs demonstrated the remarkable capability of imaging single heavy atoms (e.g., Pt and U) on low-Z substrates (Isaacson et al. 1979), as shown in Figure 22.12. These images were formed by the ADF detector collecting low-angle elastically scattered electrons only. Single atoms scatter incoherently and the image intensity is the sum of the individual atomic scattering contributions. There was sometimes a problem with thickness changes in the substrate and contributions to the ADF signal from inelastically scattered electrons. This problem was overcome by dividing the digital ADF signal by the inelastic (energy-loss) signal from the EELS system. A drawback to this technique is that diffraction contrast (e.g., from a crystalline substrate) is preserved in the low-

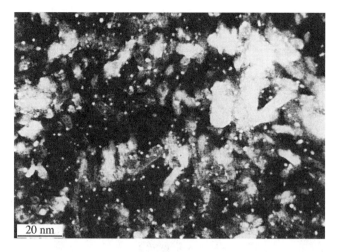

Figure 22.12. Z-contrast ADF image of individual Pt atoms or groups of atoms on a crystalline Al_2O_3 film obtained using an FEG STEM.

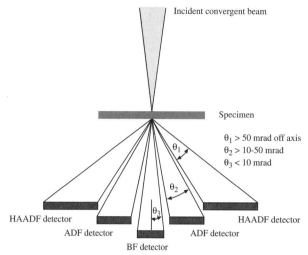

Figure 22.13. Schematic of the HAADF detector set-up for Z-contrast imaging in a STEM. The conventional ADF and BF detectors are also shown along with the range of electron scattering angles gathered by each detector.

angle EELS signal, which can confuse the image interpretation. In Figure 22.12 the large bright regions arise from the Al_2O_3 substrate diffracting onto the ADF detector, obscuring the scatter from the Pt atoms.

Because of Bragg scattering, this early approach to Z contrast was not suited to the study of crystalline specimens. Since the normal ADF detector will always collect some Bragg electrons, it was necessary to design an ADF detector with a very large central aperture. Z-contrast images could then be formed from thin crystals (Figure 22.13) (Jesson and Pennycook 1995). You can decrease the camera length with the post-specimen lenses to ensure that the Bragg electrons (including any HOLZ scatter) don't hit the detector. The image is thus formed only from the very high angle, incoherently scattered electrons.

- The detector is called a high-angle ADF or HAADF detector.
- Sometimes, the term "Howie detector" is used since Howie (1979) first proposed its design.
- Z-contrast images are also termed HAADF images.

Bragg effects are avoided if the HAADF detector only gathers electrons scattered through a semiangle of > 50 mrad (~3°). Remember that cooling your specimen has the effect of increasing coherent HOLZ scatter, so don't cool it unless you must. Electron channeling effects remain at high scattering angles, so imaging away from strong two-beam conditions and closer to zone-axis orientations is wise.

So, what do these Z-contrast images of crystals look like? Figure 22.14 shows a TEM BF image of Bi-implanted Si and below is a Z-contrast image. In the TEM BF image, formed from the direct beam, defects associated with the Bi implant are shown (we'll talk about such diffraction contrast from defects in Chapter 25) but otherwise there is no contrast associated with the Bi. In the Z-contrast image the Bi-implanted area is bright, but note that defect contrast isn't preserved in this image. You can relate the intensity differences in Figure 22.14 to an absolute measure of the Bi concentration. To do this you need to choose a suitable elastic scattering cross section. The contrast is related directly to the cross section for elastic scattering by the matrix (σ_A) and the alloying or dopant element (σ_B)

$$C = \left(\frac{\sigma_A}{\sigma_B} - F_B\right) c_B \quad [22.10]$$

where c_B is the atomic concentration of the alloying element and F_B is the fraction of the alloying element that substitutes for matrix atoms. Pennycook (1992), who pioneered this technique, found he could quantify the intensity to an absolute accuracy of ±20%.

In an FEG with probe sizes of < 0.3 nm, Z-contrast image resolution of this order is possible. Figure 22.15A shows a high-resolution phase-contrast TEM image of Ge on Si with an amorphous SiO_2 surface layer. The Si and Ge are indistinguishable by phase contrast. In Figure 22.15B, which is a STEM Z-contrast image of the same region, the higher-Z Ge crytsal region is clearly visible and the lower-Z SiO_2 layer appears very dark. The atomic structures of the Si and Ge crystals are visible in both phase-contrast and Z-contrast images, although the Z-contrast image is noisier. Phase-contrast TEM images can show similar Z-contrast effects, as we'll detail in Chapter 28. Fig-

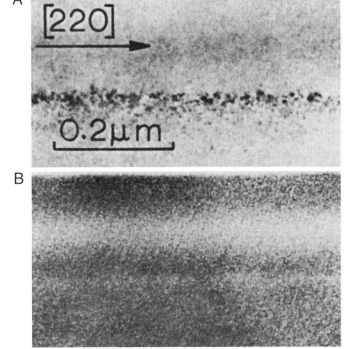

Figure 22.14. (A) Low-resolution TEM BF image showing a row of defects in Bi-implanted Si. In (B), obtained under Z-contrast conditions, the defects associated with the implant are invisible but the specimen is bright in the region implanted with Bi.

ure 22.15C shows a model of a grain boundary superimposed on a Z-contrast image which has been refined and processed to reduce the noise via a maximum-entropy approach. You can easily see atomic level detail.

> HAADF has the advantage that the contrast is generally unaffected by small changes in objective lens defocus (Δf) and specimen thickness.

We'll see in Chapter 28 that interpretation of atomic-resolution phase-contrast images requires knowledge of t and Δf. Some microscopists claim that Z contrast will become the principal method of high-resolution imaging in the future as more FEG STEMs become available; others strongly disagree.

We can think of the image in Figure 22.15B as a direct map of the $f(\theta)$ variation in the specimen. In that respect it is similar to an X-ray map showing the distribution of a certain element.

> The $f(\theta)$ map can have atomic level resolution, which XEDS imaging can't provide.

So why do we need a STEM for Z-contrast imaging? We are constrained in TEM if we use an analog screen

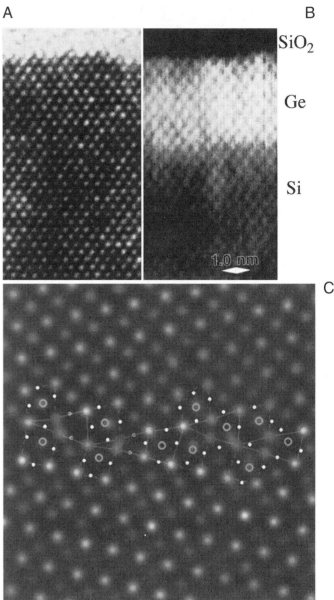

Figure 22.15. (A) High-resolution phase-contrast image of epitaxial Ge on Si with an amorphous SiO_2 surface. The bright array of dots common to the crystalline region represents atomic rows and the Ge and Si regions are indistinguishable. (B) The high-resolution Z-contrast STEM image shows the atom rows but with strong contrast at the Si–Ge interface and low intensity in the low-Z oxide. (C) Model structure of a boundary in $SrTiO_3$ superimposed on a processed Z-contrast image.

rather than a digital detector to form the image. Nevertheless, we can do Z-contrast imaging in a TEM but we have to create electron-optical conditions which are equivalent to those used in STEM. So the beam-convergence angle in TEM must equal the collection angle of the HAADF detector. This is an example of the so-called "principle of reciprocity" which we'll discuss in more detail in the next section. To converge the TEM beam to the required angular range we use so-called "hollow-cone" illumination, which requires an annular C2 aperture. However, the highest incidence angles possible in hollow-cone illumination are typically a few mrad rather than the 50–150 mrad (up to ~9°) collected by the STEM HAADF detector. So TEM Z-contrast images are not equivalent to STEM and will always contain some diffraction contrast from crystalline specimens. This leads us into the topic of diffraction contrast, which is the other form of amplitude contrast we see in TEM images.

22.5. TEM DIFFRACTION CONTRAST

Bragg diffraction, as you now know from Part II, is controlled by the crystal structure and orientation of the specimen. We can use this diffraction to create contrast in TEM images. Diffraction contrast is simply a special form of amplitude contrast because the scattering occurs at special (Bragg) angles. We've just seen how incoherent elastic scattering causes mass-thickness contrast. Now we'll see how coherent elastic scattering produces diffraction contrast. As you know, crystalline specimens usually give a single-crystal DP, such as in Figure 22.2. So, as for mass-thickness contrast, we can form BF images by placing the objective aperture round the direct beam (Figure 22.2A) and DF images come from any of the diffracted beams (Figure 22.2B). Remember that the incident electrons must be parallel in order to give sharp diffraction spots and strong diffraction contrast. So, if you can, underfocus C2 to spread the beam.

22.5.A. Two-Beam Conditions

There is one major difference between forming images to show mass-thickness contrast or diffraction contrast. We used *any* scattered electrons to form a DF image showing mass-thickness contrast. However, to get good strong diffraction contrast in both BF and DF images we tilt the specimen to *two-beam conditions*, in which only one diffracted beam is strong. Of course, the direct beam is the other strong spot in the pattern.

Remember: the electrons in the strongly excited $hk\ell$ beam have been diffracted by a *specific* set of $hk\ell$ planes and so the area that appears bright in the DF image is the area where the $hk\ell$ planes are at the Bragg condition. Hence the DF image contains *specific* orientation information, not just general scattering information as is the case for mass-thickness contrast.

We can tilt the specimen to set up several different two-beam conditions. Figure 22.16A includes a zone-axis DP from a single-crystal specimen in which the beam direction is [011]. The surrounding patterns are a series of two-beam conditions in which the specimen has been tilted slightly so that different $hk\ell$ spots are strongly excited in each pattern. We can form DF images from each strongly diffracted beam after tilting the specimen, and each will give a different image.

As you can see in Figures 22.16B, and C, BF and DF images show complementary contrast under two-beam conditions. We'll explain the image contrast in detail in Chapter 23. Obviously, to set up a series of two-beam conditions we need precise tilt control, which explains why a double-tilt eucentric holder is essential for viewing crystalline specimens.

> If you're working with crystalline materials, you'll spend a lot of time tilting the specimen to set up different two-beam conditions.

We'll see in the following chapters that two-beam conditions are not only necessary for good contrast, they also greatly simplify interpretation of the images. This is why we emphasized two-beam theory in our discussion of diffraction in Part II.

22.5.B. Setting the Deviation Parameter, s

Setting up two-beam conditions is very simple. While looking at the DP, tilt around until only one diffracted beam is strong, as in Figure 22.16. As you can see, the other diffracted beams don't disappear because of the relaxation of the Bragg conditions, but they are relatively faint. Now if you just do as we've described, the contrast might still not be the best. For reasons we'll discuss in detail in the next chapter, to get the best contrast from defects, your specimen shouldn't be exactly at the Bragg condition (**s** = 0) as in Figure 22.17A. Tilt your specimen close to the Bragg condition but make **s** small and positive (the excess $hk\ell$ Kikuchi line just outside the $hk\ell$ spot). This will give you the best possible strong-beam image contrast, as in Figure 22.17B. If you tilt the specimen slightly, so **s** increases further as shown in Figure 22.17C, the defect images become narrower but the contrast is reduced.

> Never form strong-beam images with **s** negative; the defects will be difficult to see.

22.5.C. Setting Up a Two-Beam CDF Image

We described the basic mechanism of forming BF and DF images back in Chapter 9 (Figure 9.14A). To produce the

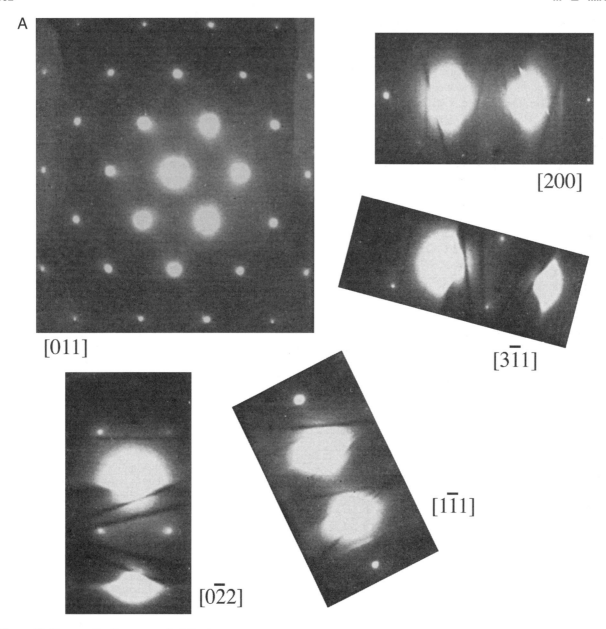

Figure 22.16. (A) The [011] zone-axis diffraction pattern has many planes diffracting with equal strength. In the smaller patterns, the specimen is tilted so there are only two strong beams, the direct 000 on-axis beam and a different one of the $hk\ell$ off-axis diffracted beams. Complementary (B) BF and (C) DF images of Al-3 wt.% Li taken under two-beam conditions are shown also. In (B) the Al$_3$Li precipitate phase (present as tiny spheres in the grain and coarse lamellae at the boundary) is diffracting strongly and appears dark. In (C), imaged with a precipitate spot, only the diffracting precipitates appear bright.

best BF diffraction contrast, tilt to the desired two-beam condition as in Figure 22.18A, and insert the objective aperture on axis as in Figure 22.2A. A two-beam CDF image is not quite as simple. You might think it involves just tilting the incident beam so the strong $hk\ell$ reflection moves onto the optic axis. If you do that, you'll find that the $hk\ell$ reflection becomes weaker as you move it onto the axis and the $3h3k3\ell$ reflection becomes strong, as shown in Figure 22.18B. What you've just done is in fact set up a *weak-beam* image condition, which we'll discuss in Chapter 26. To set up a *strong-beam* CDF image, tilt in the $\bar{h}\bar{k}\bar{\ell}$ reflection which was initially weak, and it becomes strong as it moves on axis, as shown in Figure 22.18C. The CDF technique is absolutely crucial for obtaining and interpreting diffraction-contrast images, so we will take you through it in detail.

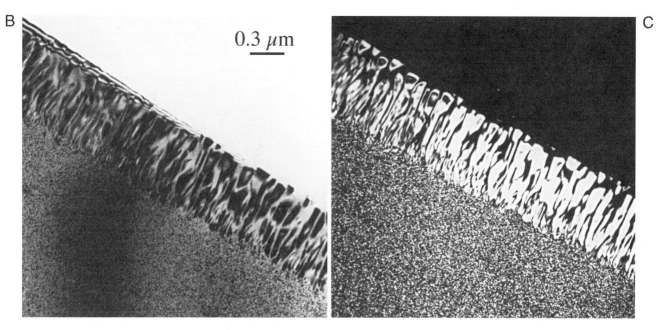

Figure 22.16. *(Continued)*

- Look at the SAD pattern and tilt the specimen until the desired $hk\ell$ reflection is strong. Make sure the incident beam is well underfocused.
- Now tilt the specimen until the $\bar{h}\bar{k}\bar{\ell}$ reflection is strong: $hk\ell$ will now be weak.
- Use the DF tilt controls to move the 000 reflection toward the strong $\bar{h}\bar{k}\bar{\ell}$ reflection. The weak $hk\ell$ reflection will move toward the optic axis and become strong.
- When $hk\ell$ is close to the axis, switch off the DF deflectors, insert and carefully center the objective aperture around 000.
- Switch the DF tilt coils on and off while looking through the binoculars. Check that the $hk\ell$ and 000 reflections appear in the same position.
- Make fine adjustments to the DF coils until you can see no shift between 000 and $hk\ell$ when the deflectors are off and on, respectively.
- Switch to image mode. If necessary, condense the beam slightly with C2 until you can see the CDF image. If you can't see an image, either the $hk\ell$ reflection is too weak (unlikely) or your tilt coils are misaligned (common). In the latter case, realign the coils (see the manufacturer's handbook).

Now go back and study Figure 9.14C carefully. You'll see that the beam was tilted through an angle $2\theta_B$ to bring the weak beam in Figure 9.14B onto the optic axis.

22.5.D. Relationship Between the Image and the Diffraction Pattern

From what we've just described, there is clearly an important relationship between the DP and a diffraction-contrast image. If we change the DP in any way, the contrast in the image will change. So it is critical to relate the DP to the image. We need to indicate the direction of the **g** vector in the image. To relate the two, remember that you may have to calibrate the rotation between the image and the DP if, whenever you change magnification, your image rotates but your DP does not. We described this calibration in Section 9.6. You should usually show the **g** vector in any BF or DF diffraction contrast image after correcting for any rotation between the image and the DP.

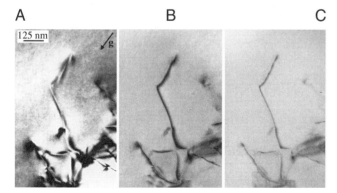

Figure 22.17. Variation in the diffraction contrast when **s** is varied from (A) zero to (B) small and positive and (C) larger and positive.

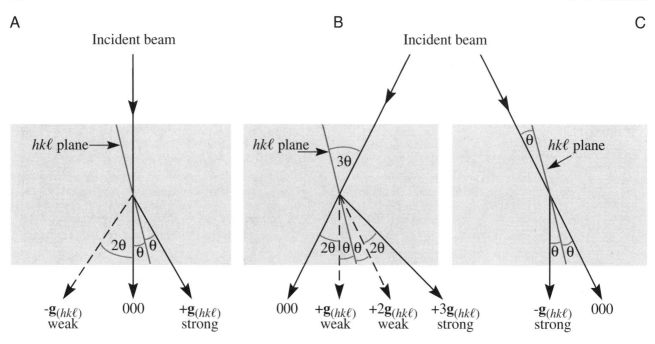

Figure 22.18. (A) Standard two-beam conditions involve the 000 spot and the $hk\ell$ spot bright because one set of $hk\ell$ planes is exactly at the Bragg condition. (B) When the incident beam is tilted through 2θ so that the excited $\mathbf{g}_{hk\ell}$ spot moves onto the optic axis, the $\mathbf{g}_{hk\ell}$ intensity decreases because the $\mathbf{g}_{3h3k3\ell}$ spot becomes strongly excited. (C) To get a strong $\bar{h}\bar{k}\bar{\ell}$ spot on axis for a CDF image, it is necessary to set up a strong $\mathbf{g}_{hk\ell}$ condition first of all, then tilt the initially weak $\mathbf{g}_{\bar{h}\bar{k}\bar{\ell}}$ spot onto the axis.

We will expand on diffraction contrast in far more detail in the subsequent chapters, making use of the fundamental operational principles we have just described.

22.6. STEM DIFFRACTION CONTRAST

The principle of forming BF and DF images in STEM is just the same as for mass-thickness contrast, i.e., use the BF detector to pick up the direct beam and the ADF detector to pick up the diffracted beams. To preserve two-beam conditions, the ADF detector must only pick up one strong diffracted beam and this can be ensured by inserting the objective aperture and selecting only one diffracted beam. Alternatively, the DP could be displaced so the chosen $hk\ell$ reflection falls on the BF detector. Either way the CRT will display a DF image.

However, the diffraction contrast observed in the STEM image will generally be much poorer than TEM contrast; the normal STEM operating conditions are not equivalent to the TEM conditions that ensure strong diffraction contrast. To understand the contrast in STEM images you need to know the beam convergence and detector collection semiangles. It's rare in fact that you'll need to do this, but we showed you how to determine the beam-convergence semiangle back in Section 5.5. To calculate the collection semiangle, you need to carry out a similar exercise as we use to determine the EELS spectrometer collection semiangle in Section 37.4.

Remember, there are three conditions that must be fulfilled for strong contrast in your image:

- The incident beam must be parallel, i.e., the convergence angle must be very small.
- The specimen must be tilted to a two-beam condition.
- Only the direct beam or the one strong diffracted beam must be collected by the objective aperture.

This condition is shown schematically in Figure 22.19A. We define the TEM convergence semiangle as α_T and the objective aperture collection semiangle as β_T. In a STEM, the equivalent angles are the beam-convergence angle α_S and the STEM detector collection angle β_S, as shown in Figure 22.19B. Therefore, we have identical operating conditions if

$$\alpha_T = \alpha_S \qquad [22.11a]$$

and

$$\beta_T = \beta_S \qquad [22.11b]$$

Now it should be immediately clear that we can't get such equivalence in a STEM because the convergence angle of

22 ■ IMAGING IN THE TEM

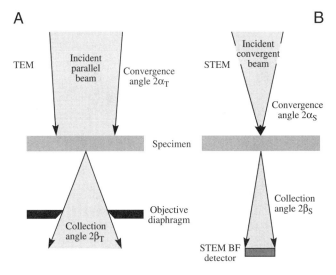

Figure 22.19. Comparison of the important beam-convergence and divergence angles (A) in TEM and (B) in STEM.

the beam is very much greater than in a TEM (since in STEM we deliberately create a convergent rather than a parallel beam). However, there is a way around this dilemma and it depends on a theorem that is often used in electron optics, called the *principle of reciprocity*. In essence, this principle says that so long as the electron ray paths contain equivalent angles (of convergence and collection) at some point in the electron optical system, the image contrast will be identical.

In other words, while the conditions in equations 22.11 can't be fulfilled, we can create conditions such that

$$\alpha_S = \beta_T \qquad [22.12a]$$

and

$$\alpha_T = \beta_S \qquad [22.12b]$$

Under these circumstances the electrons in TEM and STEM do see equivalent angular constraints, although not at the equivalent points of convergence and collection.

- Since the objective-aperture collection angle in TEM is about equal to the convergence angle in STEM, the first of this pair of equations is easily satisfied.
- To satisfy the second pair, we have to make a very small STEM collection semiangle β_S.

We can't simply increase α_T, because we must keep a parallel beam to get good TEM diffraction contrast and making the beam nonparallel (large α_T) destroys the contrast.

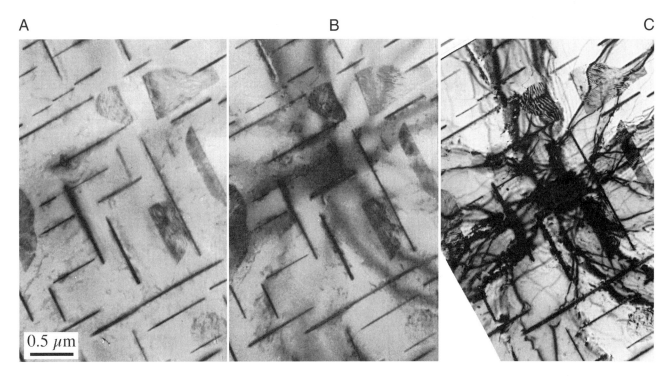

Figure 22.20. (A) BF STEM image of an Al-4 wt.% Cu specimen showing weak diffraction contrast in the form of faint bend contours. As the STEM detector collection angle is lowered (B) the diffraction contrast increases slightly at the expense of increased noise in the image. Even at the smaller collection angle, comparison with the contrast in the TEM image (C) is unfavorable. Note that the Cu-rich θ' precipitates maintain strong mass contrast in all the images.

There is an obvious drawback to making β_S small. The signal falling on the STEM detector becomes very small and the STEM image becomes noisy. So STEM diffraction-contrast images become noisier as we attempt to increase the amount of diffraction contrast, as in Figure 22.20. (See the next chapter for an explanation of the contrast (bend contours) in this figure.) Having an FEG helps to offset this increase in noise, but in general STEM diffraction-contrast images (in both BF and DF) compare so unfavorably with TEM images (see Figure 22.20C) that, while they may be useful if you're performing microanalysis, they are rarely used to show diffraction-contrast images of crystal defects. This is solely the domain of TEM, as we'll discuss in detail in the next few chapters.

CHAPTER SUMMARY

Mass-thickness contrast and diffraction contrast are two forms of amplitude contrast. Both arise because the specimen scatters electrons. The operational procedures to produce BF and DF images are identical. Interpretation of mass-thickness contrast is generally simpler than interpretation of diffraction contrast. In fact, the interpretation of diffraction contrast is sufficiently complex that we need to devote several chapters to the various forms arising in perfect and imperfect crystals.

We can summarize the characteristics of mass-thickness contrast:

- Areas of greater Z and/or t scatter electrons more strongly, and therefore appear dark in BF images and bright in DF images. The contrast can be quantified if necessary.
- TEM mass-thickness contrast images are of better quality (lower noise and higher resolution) than STEM images, but digital STEM images can be processed to show higher contrast than analog TEM images.
- STEM mass-thickness contrast images are most useful for thick and/or beam-sensitive specimens.
- Z-contrast (HAADF) images can show atomic-level resolution.

We can summarize the characteristics of diffraction contrast:

- Diffraction contrast arises when the electrons are Bragg-scattered.
- To form a diffraction-contrast image in TEM, the objective aperture selects one Bragg scattered beam. Often, the STEM detectors gather several Bragg beams which reduce diffraction contrast.
- Diffraction-contrast images in TEM always show better contrast than in STEM images, which are always noisier and almost never used.

REFERENCES

General References

Grundy, P.J. and Jones, G.A. (1976) *Electron Microscopy in the Study of Materials,* E. Arnold, Ltd., London.
Heidenreich, R.D. (1964) *Fundamentals of Transmission Electron Microscopy,* p. 31, John Wiley & Sons, New York.
Humphreys, C.J. (1979) *Introduction to Analytical Electron Microscopy* (Eds. J.J. Hren, J.I. Goldstein, and D.C. Joy), p. 305, Plenum Press, New York.
Sawyer, L.C. and Grubb, D.T. (1987) *Polymer Microscopy,* Chapman and Hall, New York.
Watt, I.M. (1985) *The Principles and Practice of Electron Microscopy,* Cambridge University Press, New York.
Williams, D.B. (1987) *Practical Analytical Electron Microscopy in Materials Science,* 2nd edition, Philips Electron Optics Publishing Group, Mahwah, New Jersey.

Specific References

Cosslett, V.E. (1979) *Phys. stat. sol.* **A55**, 545.
Howie, A. (1979) *J. Microsc.* **177**, 11.
Isaacson, M., Ohtsuki, M., and Utlaut, M. (1979) in *Introduction to Analytical Electron Microscopy* (Eds. J.J. Hren, J.I. Goldstein, and D.C. Joy), p. 343, Plenum Press, New York.
Jesson, D.E. and Pennycook, S.J. (1995) *Proc. Roy. Soc.* (London) **A449**, 273.
Pennycook, S.J. (1992) *Ann. Rev. Mat. Sci.* **22**, 171.
Reimer, L. (1993) *Transmission Electron Microscopy; Physics of Image Formation and Microanalysis,* 3rd edition, p. 194, Springer-Verlag, New York.

Thickness and Bending Effects

23

23.1. The Fundamental Ideas 369
23.2. Thickness Fringes 369
23.3. Thickness Fringes and the Diffraction Pattern 371
23.4. Bend Contours (Annoying Artifact, Useful Tool, and Invaluable Insight) 372
23.5. ZAPs and Real-Space Crystallography 373
23.6. Hillocks, Dents, or Saddles 374
23.7. Absorption Effects 374
23.8. Computer Simulation of Thickness Fringes 375
23.9. Thickness-Fringe/Bend-Contour Interactions 375
23.10. Other Effects of Bending 376

CHAPTER PREVIEW

We see diffraction contrast in an image of a perfect specimen for two reasons: either the thickness of the specimen varies or the diffraction conditions change across the specimen.

The *thickness* effect: When the thickness of the specimen is not uniform, the coupling of the direct and diffracted beams occurs over different distances, thus producing a thickness effect. Don't confuse diffraction contrast due to thickness changes with mass-thickness contrast discussed in the previous chapter. The effects are very different. The diffraction contrast changes with tilt, but the mass-thickness contrast doesn't.

The *bending* effect: Whenever the orientation of the diffracting planes changes, i.e., when the diffracting planes bend, the contrast changes. To interpret changes in image contrast we need to understand how the contrast is related to thickness and bending.

We call these two important contrast phenomena "thickness fringes" and "bend contours."
The present chapter is particularly important for three reasons:

- All TEM specimens are thin but their thickness is rarely constant.
- Because the specimens are so thin they also bend elastically, i.e., the lattice planes physically rotate.
- The planes also bend when lattice defects are introduced.

We can see the effects of these rotations even when they are < 0.1°, since they still have a significant effect on the image contrast. Therefore, the bending may arise because the specimen is thin (i.e., giving possible artifacts of the technique) or it may be caused by strains which were present in the bulk material. The result is that, in real specimens, bending and thickness effects often occur together.

Thickness and Bending Effects

23

23.1. THE FUNDAMENTAL IDEAS

To understand the origin of thickness fringes and bend contours we limit our discussion to the two-beam situation and recall equations 13.46 and 13.47, which we derived from the Howie–Whelan equations. The intensity of the Bragg-diffracted beam is then given by

$$I_g = |\phi_g|^2 = \left(\frac{\pi t}{\xi_g}\right)^2 \left(\frac{\sin^2(\pi t s_{eff})}{(\pi t s_{eff})^2}\right) = 1 - I_0 \quad [23.1]$$

where s_{eff} is the effective excitation error

$$s_{eff} = \sqrt{s^2 + \frac{1}{\xi_g^2}} \quad [23.2]$$

Although we will concentrate on I_g (the DF image intensity) for most of this discussion, the direct beam (BF image) behaves in a complementary manner (neglecting, for now, the effect of absorption). The diffracted intensity is periodic in the two independent quantities, t and s_{eff}. If we imagine the situation where t remains constant but s (and hence s_{eff}) varies locally, then we produce bend contours. Similarly, if s remains constant while t varies, then thickness fringes will result.

This chapter is simply concerned with the physical understanding of equation 23.1 and how you can relate the image to the information contained in the diffraction pattern. Although these effects are often a hindrance to systematic analysis of lattice defects, they can, in certain situations, be useful. The most important reason for understanding them is that they are unavoidable!

23.2. THICKNESS FRINGES

As a result of the way that we thin TEM specimens, very few of them (only evaporated thin films or ideal ultrami-crotomed sections) have a uniform thickness over their entire area. A BF/DF pair of images from the same region of a specimen at 300 kV is shown in Figure 23.1; the thin area is generally in the form of a wedge.

Consider again equation 23.1. You should remember that, in this calculation, t is not the "thickness" of the foil; it is actually the distance "traveled" by the diffracted beam. If we try to treat the many-beam situation rigorously, then the value of t would, in general, be different for each beam. If you are actually viewing the foil flat-on (i.e., one surface normal to the beam), then t will be close to the geometric thickness of the foil. However, it is more difficult to analyze the image thoroughly when the foil is wedge-shaped and inclined to the beam. We almost invariably make the approximation that t is fixed with the justification being that the Bragg angles are small.

Equation 23.1 tells us that the intensity of both the **0** and the **g** beams oscillates as t varies. Furthermore, these oscillations are complementary for the DF and BF images, as we show schematically in Figure 23.2. You can, of course, confirm this observation at the microscope by forming the image without using an objective aperture; there is then minimal contrast when you're in focus. The intensity, I_0, of the incident beam starts equal to unity and gradually decays, while the intensity of the diffracted beam, I_g, gradually increases until it becomes unity; I_0 is then zero; the process then repeats itself. In reality, the situation is complicated by the presence of other diffracted beams (we are never truly in a two-beam situation) and absorption.

> As a rule of thumb, when other diffracted beams are present the effective extinction distance is reduced. At greater thicknesses, absorption occurs and the contrast is reduced.

These oscillations in I_0 or I_g are known as thickness fringes, though they are often not fringes. We sometimes call them thickness contours, because they denote the con-

Figure 23.1. (A) DF and (B) BF images from the same region of a wedge-shaped specimen of Si at 300 kV tilted so that **g**(220) is strong. The periodicity and contrast of the fringes is similar and complementary in each image.

tours where the specimen has constant thickness; you will only see these fringes when the thickness of the specimen varies locally, otherwise the contrast will be a uniform gray. As we'll see, the actual contrast can quickly change if the specimen is tilted through a small angle.

It is important to realize that the image may appear to be black or white depending on the thickness of the specimen.

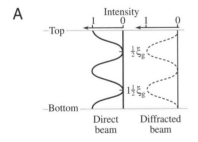

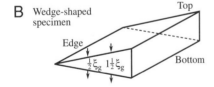

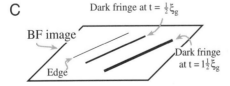

Figure 23.2. (A) At the Bragg condition (s = 0), the intensities of the direct and diffracted beams oscillate in a complementary way. (B) For a wedge specimen, the separation of the fringes in the image (C) is determined by the angle of the wedge and the extinction distance, ξ_g.

For example, in BF images, thicker areas are often brighter than thinner areas, which really is counterintuitive.

Several examples of how thickness fringes might appear in your image are shown in Figure 23.3. Although it is often helpful to think of these fringes as thickness con-

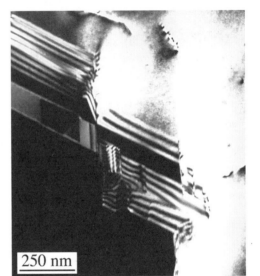

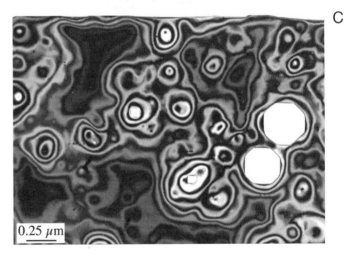

Figure 23.3. Examples of thickness fringes in (A) DF image of a preferentially thinned grain boundary; (B) a strong 220 DF image of microtwinned GaAs taken with only the left-hand grain diffracting, and (C) BF image of a chemically etched thin film of MgO. The white regions in (C) are holes in the specimen.

23 ■ THICKNESS AND BENDING EFFECTS

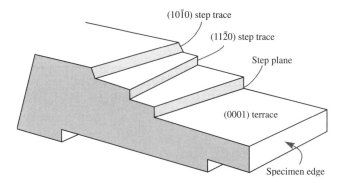

Figure 23.4. Schematic cross-sectional view of a specimen with terraces parallel to the surface and steps connecting terraces.

tours analogous to height or depth contours on a map, with the hole at sea level, remember that there are *two* surfaces to your TEM specimen. A DF image will usually appear to give greater contrast. This is partly because any hole now appears dark, but also because many-beam effects are less important in DF. In Figure 23.3A the narrow fringe pattern in this DF image is due to the grain boundary region being thinner than the matrix. In the DF image in Figure 23.3B, the reflection used to form the image is only excited in the right grain so the left grain is black; the diffracting grain exhibits strong thickness fringes in the regions where there are microtwins. This image introduces the idea that images of defects can also show thickness effects.

In Figure 23.3C, the specimen is an almost flat, parallel-sided film of MgO with holes formed by chemical etching; after thinning, the surfaces were faceted by heating the specimen at 1400°C. The holes in the image are white, so it is a BF image. The contours, like the holes, are angular because of the faceting, but they are not uniformly spaced because it's not a uniform wedge. Notice that the center of the hole is faceted and that the first fringe is a very narrow dark line. We know that this surface is different because it is curved. We also know from Figure 23.2 that, in a BF image, the first fringe must be bright if the thickness actually decreases to zero. We can therefore conclude from this one image that the specimen is not tapering to zero thickness at the center of this hole.

Although we've talked about wedges or specimens with gently curving surfaces so far, the way we actually calculate and analyze the contrast from such wedges is shown in Figure 23.4. We imagine that the specimen has two parallel surfaces which are normal to the electron beam, so that we have a fixed thickness for each calculation. We also assume that the beam is normal to the surface. We then change t and recalculate the intensity. Finally, we plot the different values for the intensity against t but we never actually inclined the surfaces!

Figure 23.5 shows a striking image of a wedge-shaped specimen of Al_2O_3 which has been heat treated so that the surface has facets parallel to certain low-index planes. The thickness fringes can then be seen to be discrete regions of different shades of gray; the fringes are, in general, quantized. You can form similar specimens by cleaving layer materials (e.g., graphite) but the specimens tend to bend, which obscures these abrupt contrast changes.

23.3. THICKNESS FRINGES AND THE DIFFRACTION PATTERN

A general rule in TEM is that, whenever we see a periodicity in real space (i.e., the image), there must be a corresponding array of spots in reciprocal space; the converse is also true. If we image a specimen with a constant wedge angle, then we will see a uniform spacing of thickness fringes in both the BF and DF two-beam images even when $s_g = 0$. We must therefore have more than one spot "at G" when $s_g = 0$, otherwise we would not see fringes. We already know that if we increase s or if the wedge angle were larger, then the fringe separation would decrease and the spacing of these spots must therefore increase.

To understand why there is more than one spot at G, go back to Chapter 17 where we showed that, because the specimen is thin, any spot in the DP will be elongated normal to the surface. When the specimen is wedge-shaped, there will be two surfaces and we can imagine the spot being elongated normal to both surfaces, as was shown in Figure 17.4. Actually, we have two curved relrods which do not intersect at $s = 0$; we related this curvature to the dispersion surface in Chapter 15. The diffraction geometry close to G is shown in Figure 23.6.

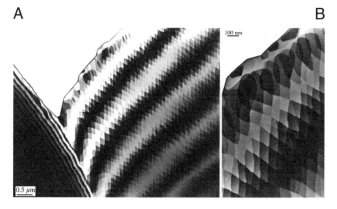

Figure 23.5. Thick fringes from an annealed Al_2O_3 specimen with the geometry shown in Figure 23.4. (A) At low magnification, the fringes are well defined and continuous, even when the wedge angle and wedge axis change. (B) At higher magnification, the contrast is seen to be quantized within a given fringe.

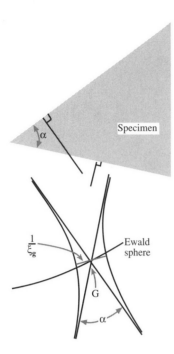

Figure. 23.6. Relrods aligned normal to both surfaces of a wedge-shaped specimen. In practice, the relrods don't cross so there are always two spots in the DP.

> The minimum spot spacing in the DP corresponds to the periodicity of the thickness fringes, which at **s** = 0 is given directly by the extinction distance.

This spot spacing is thus related to ξ_g^{-1}. While the spacing of the thickness fringes depends on ξ_g, it is not equal to ξ_g. As we referenced back in Chapter 17, Amelinckx's group has shown that we can describe this geometric relationship as shown in Figure 17.4. As we tilt the crystal away from **s** = 0, the Ewald sphere will move up or down (as **s** becomes negative or positive) to cut the two "rods." So, you can see that there will be two spots instead of one at G and their separation will increase as **s** increases. As the separation increases, the spacing of the fringes decreases; the thickness fringes move closer together because the ξ_{eff} has decreased. The change in the fringe spacing is similar either side of **s** = 0.

> So be wary of trying to make accurate thickness measurements in wedge-shaped crystals.

We refer to thickness fringes as being an example of amplitude contrast because, in the two-beam case, they are associated with a particular reflection, **g**. They actually occur due to interference between two beams, both of which are located close to **g**, so they are really an example of phase contrast although we rarely think of them as such.

23.4. BEND CONTOURS (ANNOYING ARTIFACT, USEFUL TOOL, AND INVALUABLE INSIGHT)

This is a particularly satisfying topic, because you can understand it by considering a simple physical picture and yet the concept involved is the basis for understanding most aspects of defect contrast. Bend contours (don't call them extinction contours) occur when a particular set of diffracting planes is not parallel everywhere; the planes rock into, and through, the Bragg condition.

The specimen shown schematically in Figure 23.7 is aligned so that the $hk\ell$ planes are exactly parallel to the incident beam at the center of the figure and always lie normal to the specimen surface even when it bends. We imagine that the foil bends evenly, so that the $hk\ell$ planes are exactly in the Bragg condition at A and the $\bar{h}k\bar{\ell}$ planes are exactly in the Bragg condition at B. We can draw the systematic row of reflections as you see below the bent crystal. Notice that –G is now on the left and G on the right. Now if we form a BF image we will see two dark lines. Next, we form the DF image using reflection **g**. We see a

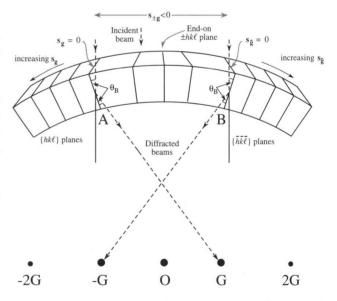

Figure. 23.7. The origin of bend contours shown for a foil symmetrically bent either side of the Bragg conditions. For this geometry, when the $hk\ell$ planes are in the Bragg condition, the reflection G is excited. Notice that G and the diffracting region are on opposite sides of O; if the foil were bent upwards, they would be on the same side.

bright band on the left because that's where **g** is excited. Now use **ḡ** to form the image and the bright band is on the right. These bands are referred to as bend contours. Experimental images are shown in Figures 23.8 and 23.9.

> In actually doing this imaging experiment you should translate the objective aperture to form the DF images; you'll lose some resolution but don't move the specimen.

Remembering Bragg's law, the ($2h\ 2k\ 2\ell$) planes diffract strongly when θ has increased to $\sim 2\theta_B$. So we'll see extra contours because of the higher-order diffraction. As θ increases, the planes rotate through the Bragg condition more quickly (within a small distance Δx) so the bend contours become much narrower for higher-order reflections.

Tyro-microscopists occasionally have difficulty in distinguishing higher-order bend contours from real line defects in the crystal. The solution is very simple: tilt your specimen. Bend contours are not fixed to any particular position in the specimen and move as you tilt.

Bend contours are true amplitude contrast, not phase contrast.

23.5. ZAPS AND REAL-SPACE CRYSTALLOGRAPHY

In the above discussion we only considered bending about one axis. In real specimens, the bending will be more complex. This complexity will be important when the bent area is oriented close to a low-index pole, because the bend contours form a zone-axis pattern, or ZAP. Two examples of these ZAPs are shown in Figure 23.8. Although the ZAP is distorted, the symmetry of the zone axis is clear and such patterns have been used as a tool for real-space crystallographic analysis (e.g., Rackham and Eades 1977). Each contour is uniquely related to a particular set of diffracting planes, so the ZAP does not automatically introduce the twofold rotation axis that we are used to in SAD patterns. These contours are the real-space analog of the symmetry in large-angle CBED patterns.

In fact, it's the exception that a ±**g** pair of bend contours are straight and parallel. In case you are having a problem visualizing how a pair of contours might diverge, go back to the bent specimen in Figure 23.7, hold the $\bar{h}\bar{k}\bar{\ell}$ plane fixed at $x = x_0$, and then as you move along the foil (going into the page) gradually decrease the bend in the foil. The position where the $hk\ell$ planes are in the Bragg condition gradually moves to the left, so $-x_0$ becomes more

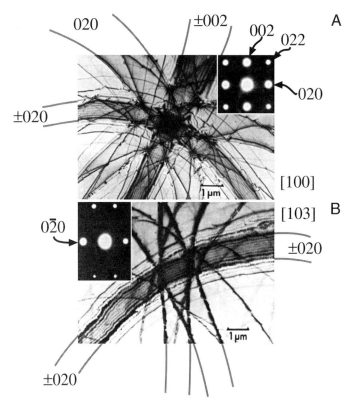

Figure. 23.8. BF images of a bent Al specimen oriented close to the (A) [100] and (B) [103] zone axes. These images are known as (real-space) ZAPs, or zone-axis patterns, and are shown with their respective zone-axis diffraction patterns (insets). Each diffracting plane produces two bend contours, depending on whether θ_B or $-\theta_B$ is satisfied. Note that the separation of the bend contours is not uniform for any particular pair of planes because the curvature of the bending is not, in general, the same.

negative. Since x_0 is fixed the contours move apart in the image.

Notice how at the zone axis, the main 200 contours in the [100] ZAP are closely spaced, while in the [103] pattern, only one of contours is more closely spaced; the others are more clearly defined and further apart.

> In this case, small **g** in the DP gives a small spacing in the image, contrary to the usual inverse relationship between image and DP.

When the foil curvature is equal, this effect allows you to recognize a low-index ZAP. Since you can tilt the crystal you can form different ZAPs from exactly the same area of your specimen just as you can for SAD and CBED (where we also use the term ZAP but then it refers to a DP). You index the contours in the manner described in Section 18.4 but use all the spots in the zone-axis SAD pattern.

If your specimen is buckled, you can tilt it so that a particular bend contour stays at the position you're studying in the image. You're then doing the same operation as we did using Kikuchi lines in Chapter 19. Tilting in image mode is more tricky, but if the specimen is very buckled or too thin, you can't use Kikuchi lines. The ZAP and bend contours let you work in real space. You can even set the value of **s** for a particular **g** at a particular location on your specimen!

23.6. HILLOCKS, DENTS, OR SADDLES

The simplest use for bend contours is in determining whether an area is a hillock or a dent. This information is useful when analyzing particles grown on a substrate, particularly if the substrate is a thin film.

Figures 23.9A and B show a ZAP in a thin specimen of Al_2O_3 and the associated SAD pattern; the dark bands are $\{3\bar{3}00\}$ bend contours. Figures 23.9C–H show DF images recorded using each of these reflections; the area is identical in each of these images. Using Figures 23.7 you can determine the sense of the bending. You can see directly that the bend contour from one set of $hk\ell$ planes does not necessarily lie parallel to those planes but instead both curves and changes in width.

Remember, for these contrast experiments only, it is important that you move the objective aperture, not the specimen. The resolution of the image will be lower, but this is not critical for this application.

23.7. ABSORPTION EFFECTS

When your specimen is very thick you won't see an image, so we can say that the electrons have then all been absorbed. The absorption process is more important than this obvious statement might imply. Much of our thinking about this topic is, however, just as empirical. In fact, it is common to define an imaginary component ξ'_g to the extinction distance, so

$$\xi_g^{abs} = \xi_g \left(\frac{\xi'_g}{\xi'_g + i\xi_g} \right) \qquad [23.3]$$

in the Howie–Whelan equations. The term ξ'_g is found to be approximately $0.1\xi_g$. The reason for choosing this expression for ξ_g^{abs} is that the $1/\xi_g$ in the Howie–Whelan equation is replaced by $(i/\xi'_g + 1/\xi_g)$. We do the same for ξ_0. The result is that γ in the Howie–Whelan equations has an imaginary component. Consequently, we now have an exponen-

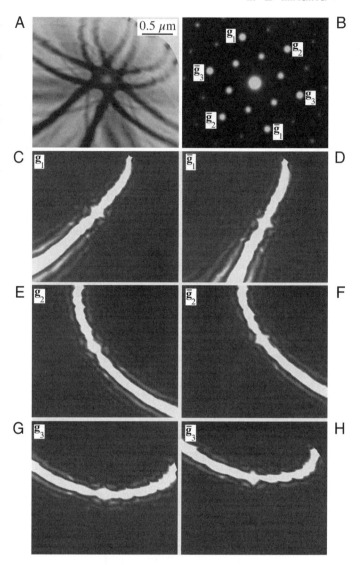

Figure 23.9. An 0001 real space ZAP of Al_2O_3: (A) BF image and (B) corresponding DP. (C–H) Displaced-aperture DF images taken from the spots indicated in (B) identifying the principal dark bend contours in (A); (C,D) ±($\bar{3}030$), (E,F), ±($\bar{3}300$), (G,H) ±($03\bar{3}0$). Note in (A) that the inner $\{11\bar{2}0\}$ spots produce fainter bend contours than the $\{\bar{3}300\}$.

tial decay of the diffracted amplitude. It's a completely phenomenological treatment, but you will see reference to it. When we discuss EELS in Part IV, you'll appreciate the difficulties in modeling the effects of inelastic scattering on the image by a single parameter.

We did briefly discuss absorption of Bloch waves in Chapter 14. We showed that Bloch wave 2 (smaller **k**) is less strongly absorbed than Bloch wave 1; Bloch wave 1 travels along the atom nuclei while Bloch wave 2 channels between them. As the crystal becomes thicker we lose

23 ■ THICKNESS AND BENDING EFFECTS

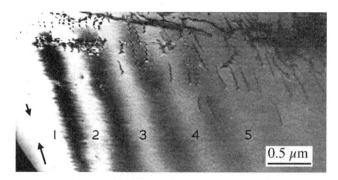

Figure. 23.10. The contrast of thickness fringes in a two-beam BF image decreases when the effect of anomalous absorption is included. Note that the defects are still visible when the fringes have disappeared at a thickness of ~5 ξ_g.

Bloch wave 1. Since thickness fringes result from a beating between the two waves, we will lose the thickness fringes but will still be able to "see through" the specimen, as you can appreciate from Figure 23.10.

> Absorption due to the loss of Bloch wave 1 is called anomalous absorption for historical reasons, not because it is unexpected.

Bend contours in thicker parts of the specimen will also show the effect of this anomalous absorption. Looking back at Chapter 15 on the dispersion surface, you'll see that when s_g is negative the tie line D_1D_2 would be closer to **0** than **g**, and Bloch wave 2 contributes to ϕ_g more strongly. When s_g becomes positive, Bloch wave 1 is the more strongly excited. We lose Bloch wave 1 because the specimen is thick. So as we rock through the Bragg condition on the bend contour, we'll lose the thickness fringes faster where Bloch wave 1 was weaker already, i.e., when s_g is negative or inside the ±**g** pair of bend contours.

We can summarize this brief description of absorption with some conclusions:

- We can define a parameter ξ'_g which is usually about $0.1\xi_g$ and is really a fudge factor that modifies the Howie–Whelan equations to fit the experimental observations.
- The different Bloch waves are scattered differently. If they don't contribute to the image, we say that they were absorbed. We thus have anomalous absorption which is quite normal!
- Usable thicknesses are limited to about $5\xi_g$, but you can optimize this if you channel the less-absorbed Bloch wave.

23.8. COMPUTER SIMULATION OF THICKNESS FRINGES

The simulation of thickness fringes can be carried out using the Head *et al.* programs or with Comis (see Section 1.5). We'll talk more about these programs in Chapter 25, but be wary: don't use a program as a "black box." Why do we need to simulate thickness fringes? As an illustration, let's look at a 90°-wedge specimen (see Section 10.6) so that we know how the thickness changes with position. The actual thickness will be *very* sensitive to the orientation of the specimen since the specimen is so thick, as shown in Figure 23.11A. The specimen is a GaAs/Al$_x$Ga$_{1-x}$As layered composite grown on (001). Since the cleavage surface it {110} it can be mounted at 45° so the beam is nearly parallel to the [100] pole. The value of ξ_g is different for the two materials, so they can be readily distinguished. Clearly, a quantitative simulation of this situation is non-trivial, especially if you also have to consider the effect shown in Figure 23.6.

Because the fringe spacing changes as ξ_g changes, it will also change if you vary the accelerating voltage. You can see this effect clearly in Figures 23.11B and C which compare the same region of a wedge specimen imaged at 300 kV and 100 kV.

23.9. THICKNESS-FRINGE/ BEND-CONTOUR INTERACTIONS

It's clear from equation 23.1 that both bending and thickness effects can occur together. This combined effect is shown in Figure 23.12, where the axis of bending runs normal to the edge of the wedge specimen. When **s** = 0, the value of ξ_{eff} is largest. As we bend away from the Bragg condition, on either side, ξ_{eff} decreases so the thickness contours curve toward the edge of the specimen. This image actually shows the **g**(111) and ($\bar{\mathbf{g}}$)($\bar{1}\bar{1}\bar{1}$) contours (arrowed). As an exercise you can calculate the value of **s** at any point between the contours in this image. Assume the wedge angle is constant and $t = 0$ at the edge; then compare the thickness you deduce using ξ_{eff} with the thickness value extrapolated from the regions where **s** = 0.

If a defect causes the specimen to bend, then the contrast from the defect and that from thickness variations will be linked.

Since the effective thickness is s_{eff}^{-1}, it will change as you increase the deviation parameter, **s**. You can use this fact to determine the thickness of an area quite accurately, providing you have a reference value such as zero thick-

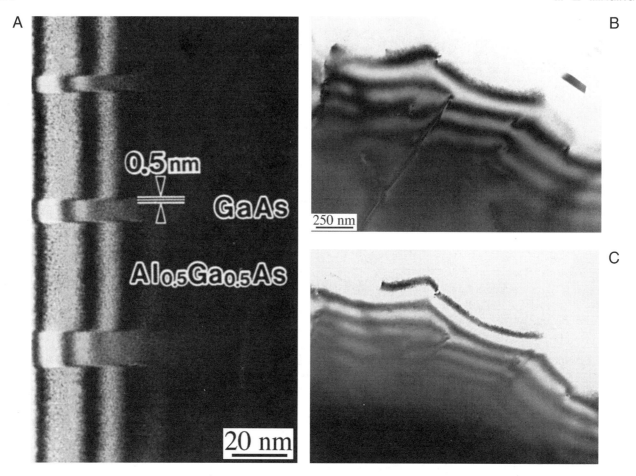

Figure 23.11. (A) Thickness fringes in a 90° wedge of alternating GaAs and AlGaAs. The extinction distance changes in each phase so the fringe spacing changes. Strong beam BF images ($s = 0$) for (B) 300-kV and (C) 100-kV electrons. The extinction distance increases as the accelerating voltage increases, and you can see through thicker areas; compare with the images in Figure 23.1.

ness at a hole in the specimen. You initially tilt the specimen so that the chosen reflection is at the Bragg condition. (Remember that this analysis assumes that you only have two beams.) You can determine s_g quite accurately if you can see the Kikuchi lines. Then you can determine s_{eff}^{-1} at different positions on the specimen. The maximum value of s_{eff}^{-1} is ξ_g and occurs at $s_g = 0$. As you tilt the specimen to increase s_g in the positive or negative sense, you'll see the thickness fringes move closer together. We'll examine the situation where s_g is very large in Chapter 26, and where the foil also bends in Section 23.11.

Aside: *You should be careful when using this method for thickness determination in XEDS analysis, since only the thickness of the diffracting (crystalline) material is determined. There may be amorphous material on the surface which has similar or different composition.*

23.10. OTHER EFFECTS OF BENDING

In some situations, the bending of the foil may be more subtle. For example, strains in TEM specimens may relax at the surface of the thin specimen. A particularly important example of this effect was found by Gibson *et al.* (1985) in the study of superlattices in semiconductors.

We'll generalize the situation a little. Imagine that two cubic materials, which normally have slightly different lattice parameters, are grown on one another to form an artificial superlattice with a (001) (i.e., cube-on-cube) interface plane. One crystal must expand and the other contract normal to this interface. When we prepare a cross-section TEM specimen, we might then imagine it relaxing at the surface as shown in Figure 23.13B. The reason for the relaxation is simply that this allows the one material to expand while the other contracts; the constraint at the surface has been removed during the specimen preparation pro-

23 ■ THICKNESS AND BENDING EFFECTS

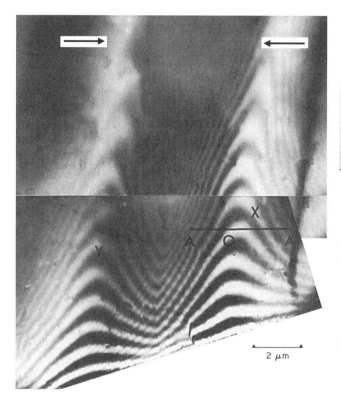

Figure 23.12. Since both thickness fringes and bend contours (X and Y) affect the contrast seen in the image, and both can occur in the same part of the specimen, they can affect, or couple with, one another to give the complex contrast shown in this BF image. Along the line A-A, **s** changes in sign, being approximately zero at O and negative between the contours.

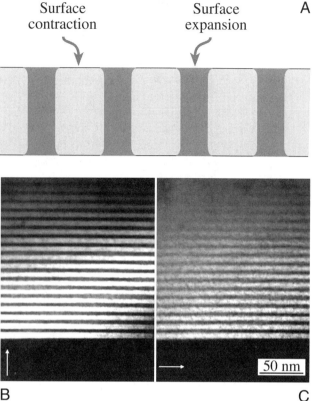

Figure 23.13. (A) A schematic of how interfaces might relax at the surface of a thin specimen. (B,C) DF images of a GaAs/AlGaAs superlattice imaged in two orthogonal reflections, 200 and 020, with the specimen oriented at the 001 pole. (B) The [020] vector is parallel to the interface while (C) the [200] is normal to it. If planes parallel to the interface bend to relax the strain caused by the lattice misfit, then only the 020 image will be affected, giving a more abrupt contrast change.

cess. This argument is admittedly crude, but Figures 23.13B and C show that images recorded with **g** = 020 normal to this interface appear sharper than images formed when **g** = 200 is parallel to the interface.

So, no matter whether **g** is 020 or 0$\bar{2}$0, the Bragg planes are bent closer to **s** = 0 at one surface or the other.

Here, bending only occurs within a short distance of the interface but it significantly affects the appearance of the DF image. The bending is actually making the image appear sharper than it should.

This example is special but emphasizes the point: relaxation at the surface can cause the specimen to bend and this bending will affect the appearance of your image.

CHAPTER SUMMARY

The effects of changes in thickness and specimen bending are both explained by equation 23.1. Although this equation was derived for a two-beam geometry, you'll see similar effects when more strongly excited beams are present but the simple $\sin^2$ dependence will be lost.

- Varying *t* while keeping **s** constant gives thickness fringes.
- Varying **s** while keeping *t* constant gives bend contours.

Thickness fringes are an interference effect and, with care, can be used to calculate the foil thickness and reveal the topography.

> Note that if the two surfaces of the specimen are parallel, then we don't see thickness fringes, even if the specimen is tilted. However, the contrast of that region will depend on the projected thickness.

Bend contours are very useful because they map out the value of **s** in the specimen. If your foil is bent around more than one axis, bend contours combine to produce beautiful ZAPs which reflect the true symmetry of the material.

However, if you want to keep defect analysis simple, then you need to avoid bending and work in relatively thin regions of nearly constant thickness. The exception to this rule is that there are quite a few special cases where you want to do exactly the opposite! So the message is that bending and thickness variations give you extra parameters which you can use in your study as long as you can control these parameters. This control comes from mastering the BF/DF/SAD techniques in Chapter 9.

Lastly, be aware that anomalous absorption is not anomalous. It can best be explained by Bloch-wave interactions.

REFERENCES

Specific References

Gibson, J.M., Hull, R., Bean, J.C., and Treacy, M.M.J. (1985) *Appl. Phys. Lett.* **46**, 649.

Rackham, G.M. and Eades, J.A. (1977) *Optik* **47**, 227.

Susnitzky, D.W. and Carter, C.B. (1992) *J. Am. Ceram. Soc.* **75**, 2463. An overview of the surface morphology of heat-treated ceramic thin films studied using TEM.

Planar Defects

24

24.1. Translations and Rotations	381
24.2. Why Do Translations Produce Contrast?	382
24.3. The Scattering Matrix	383
24.4. Using the Scattering Matrix	384
24.5. Stacking Faults in fcc Materials	385
24.5.A. Why fcc Materials?	386
24.5.B. Some Rules	386
24.5.C. Intensity Calculations	387
24.5.D. Overlapping Faults	388
24.6. Other Translations: π and δ Fringes	389
24.7. Phase Boundaries	391
24.8. Rotation Boundaries	391
24.9. Diffraction Patterns and Dispersion Surfaces	391
24.10. Bloch Waves and BF/DF Image Pairs	393
24.11. Computer Modeling	394
24.12. The Generalized Cross Section	395
24.13. Quantitative Imaging	396
24.13.A. Theoretical Basis and Parameters	396
24.13.B. Apparent Extinction Distance	397
24.13.C. Avoiding the Column Approximation	397
24.13.D. The User Interface	398

CHAPTER PREVIEW

Internal interfaces (grain boundaries, phase boundaries, stacking faults) or external interfaces (i.e., surfaces) are perhaps the most important defects in crystalline engineering materials. Their common feature is that we can usually think of them as all being two-dimensional, or planar, defects. The main topics of this chapter will be:

- Characterizing which type of internal interface we have and determining its main parameters.
- Identifying lattice translations at these interfaces from the appearance of the diffraction-contrast images.

Rotations are usually associated with line defects and they will be discussed in Chapter 25. We can't usually identify the details of the local structure of an interface unless we use HRTEM, so we will return to that topic in Chapter 28.

Planar Defects

24.1. TRANSLATIONS AND ROTATIONS

An interface is simply a surface which separates any two distinct regions of the microstructure. For most of our discussion, we will assume that the surface is flat and is thus a planar defect. We can sketch a general interface as shown in Figure 24.1.

> The upper crystal is held fixed while the lower one is translated by a vector $\mathbf{R(r)}$ and/or rotated through some angle θ about any axis, $\mathbf{v}$.

With this general definition, we can summarize the different classes of planar defects:

- *Translation boundary, RB*. Any translation $\mathbf{R(r)}$ is allowed, θ is zero, and both regions are identical and thus perfectly aligned. Stacking faults (SFs) are a special case. We'll denote the translation boundary as RB so as to avoid confusing it with the twin boundary (TB).
- *Grain boundary, GB*. Any values of $\mathbf{R(r)}$, $\mathbf{n}$, and θ are allowed, but the chemistry and structure of the two grains must be the same. The SF is again a special case, but this class also includes TBs.
- *Phase boundary, PB*. As for a GB, but the chemistry and/or structure of the two regions can differ.
- *Surface*. A special case of a PB where one phase is vacuum or gas.

Now with each of these groups, we can have special examples. We list some of the most common examples in Table 24.1, including those that we will consider in this chapter. For a more detailed discussion, we refer you to the general references at the end of the chapter.

RBs include the familiar SFs found in fcc, hcp, diamond-cubic, and layer materials. They have been widely studied because they play an important role in the mechanical properties of the fcc metals, e.g., Cu and stainless steel. They are also found in more complex materials such as spinels, Ni_3Al, Ti_3Al, etc., where the lattice parameters, and therefore the dislocation Burgers vectors, are large.

For example, the anti-phase boundary (APB) in ordered CuAu (which we can describe as two interpenetrating simple-cubic superlattices) is produced by translating one superlattice by ½<111> with respect to the other. It is called an APB because one superlattice is out of phase with the other. If the crystal were disordered and the Cu and Au occupied the bcc sites randomly, then ½<111> would be a lattice vector and no defect would exist. This particular APB can thus be regarded as an SF. Although we know that {111} is the favored SF plane for fcc metals, SFs in other materials lie on different planes. We will find that the methods used to characterize RBs can often be used to determine $\mathbf{R(r)}$ in other interfaces.

GBs fall into two groups, low-angle and high-angle. Low-angle boundaries necessarily involve a rotation which is usually accommodated by arrays of dislocations; we'll consider these defects in Chapter 25. High-angle boundaries can adopt some special values of $\mathbf{n}$ and θ such that a large fraction of lattice sites in one grain is shared by the other grain. We characterize the fraction by its inverse, which we call Σ. For example, the common twin boundary in fcc metals is the $\Sigma = 3$ grain boundary. The reason this is important to our discussion is that if a set of lattice points is common to two grains (as implied by the Σ coincident-site lattice concept), then certain planes may also be common and may give rise to common reflections. These reflections will remain common even if one grain is translated relative to the other. In that case, we'll have a special type of RB, called the rigid-body translation. Rigid-body translations in grain boundaries behave just like other SFs except $\mathbf{R}$ is usually small and is not directly related to the lattice parameters.

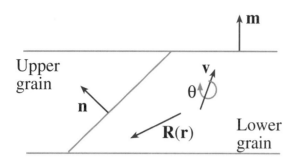

Figure 24.1. A specimen containing a planar defect. The lower grain is translated by a vector **R(r)** and rotated through an angle θ about the vector **v**, relative to the upper grain. The defect plane is **n**, the foil normal is **m**.

There is a second group of APBs where the two grains cannot be related by a translation. These occur in GaAs, ZnO, AlN, and SiC, for example. One lattice can always be related to the other by a rotation of 180° to give the equivalent of an inversion; they are sometimes known as inversion domain boundaries (IDBs). These special interfaces can often be imaged because of the small associated translation. We analyze this translation as if it were a simple RB because all the planes on one side of the IDB are parallel to their counterparts on the other; the ($hk\ell$) plane on one side is parallel to the $\bar{h}\bar{k}\bar{\ell}$ on the other. We can't distinguish **g** from **ḡ** unless we use CBED (Taftø and Spence 1982).

PBs are rarely analyzed fully. If the orientation, chemistry, and structure can all change on crossing the boundary, then not only will the reflections change, but all extinction distances will change too. Some special examples of such interfaces are hcp-Co/fcc-Co, bcc-Fe/fcc-Fe, $NiO/NiFe_2O_4$, and $GaAs/Al_xGa_{1-x}As$. Of course, the number of other such interfaces is countless.

Surface studies using TEM have quite recently become very important although the experimental tools have been available for some time. We will discuss surfaces in this chapter insofar as they are imaged by diffraction contrast. So-called profile imaging will follow in Chapter 28 on HRTEM. The two surface-sensitive techniques are plan-view and reflection electron microscopy (REM; see Chapter 31).

24.2. WHY DO TRANSLATIONS PRODUCE CONTRAST?

As usual, we will start our analysis considering only two beams, O and G. Our approach will use hand-waving arguments, which are not perfect, to justify adapting the Howie–Whelan equations for specimens containing interfaces. We'll use the same approach for other defects in Chapter 25. Because the Howie–Whelan equations for perfect crystals assume two-beam conditions, we are able to solve them analytically. We'd like to be able to do the same when defects are present, because this gives us a physical understanding of the processes which produce the contrast. There are two important features which we will need to keep in mind:

- Diffraction contrast only occurs because we have Bloch waves in the crystal. However, our analysis will initially only consider diffracted beams.
- We make the column approximation so we can solve the equations; we must be wary whenever the specimen or the diffraction conditions change within a distance comparable to the column diameter.

A unit cell in a strained crystal will be displaced from its perfect-crystal position so that it is located at position $\mathbf{r}'_n$ instead of $\mathbf{r}_n$, where n is included to remind us that we are considering scattering from an array of unit cells; we'll soon omit the n.

Table 24.1. Examples of Internal Planar Defects

Group	Structure	Example	Example
SF	Diamond-cubic, fcc, zinc blende	Cu, Ag, Si, GaAs	**R** = ⅓ [111] or **R** = ⅙ [11$\bar{2}$]
APB/IDB	Zinc blende, wurtzite	GaAs, AlN	inversion
APB	CsCl	NiAl	**R** = ½ [111]
APB/SF	Spinel	$MgAl_2O_4$	**R** = ¼ [110]
GB	All materials	Often denoted by Σ where $Σ^{-1}$ is the fraction of coincident lattice sites	rotation plus **R**
PB	Any two different materials	Sometimes denoted by $Σ_1, Σ_2$, which are not equal	rotation plus **R** plus misfit

24 ■ PLANAR DEFECTS

$$\mathbf{r}'_n = \mathbf{r}_n + \mathbf{R}_n \qquad [24.1]$$

In this expression, $\mathbf{R}_n$ is actually $\mathbf{R}_n(\mathbf{r})$; it can vary throughout the specimen. The term $e^{2\pi i \mathbf{K} \cdot \mathbf{r}}$ in equation 13.3 now becomes $e^{2\pi i \mathbf{K} \cdot \mathbf{r}'}$ so we need to examine the term $\mathbf{K} \cdot \mathbf{r}'$. We know that $\mathbf{K}$ is $\mathbf{g} + \mathbf{s}$, so we can write

$$\begin{aligned}\mathbf{K} \cdot \mathbf{r}'_n &= (\mathbf{g} + \mathbf{s}) \cdot (\mathbf{r}_n + \mathbf{R}_n) \\ &= \mathbf{g} \cdot \mathbf{r}_n + \mathbf{g} \cdot \mathbf{R}_n + \mathbf{s} \cdot \mathbf{r}_n + \mathbf{s} \cdot \mathbf{R}_n\end{aligned} \qquad [24.2]$$

Now, since $\mathbf{r}_n$ is a lattice vector, $\mathbf{g} \cdot \mathbf{r}_n$ is an integer as usual. The third term, $\mathbf{s} \cdot \mathbf{r}_n$, gives our usual sz term, so the new terms are $\mathbf{g} \cdot \mathbf{R}_n$ and $\mathbf{s} \cdot \mathbf{R}_n$.

When we discuss strong-beam images we know that $\mathbf{s}$ is very small. Since we are using elasticity theory, $\mathbf{R}_n$ must be small. Hence we ignore the term $\mathbf{s} \cdot \mathbf{R}_n$. Remember that we have made a special assumption which may not be valid in two situations:

- When $\mathbf{s}$ is large; we'll encounter this when we discuss the weak-beam technique in Chapter 26.
- When the lattice distortion, $\mathbf{R}$, is large; this occurs close to the cores of some defects.

We now modify equation 13.8 intuitively to include the effect of adding a displacement from equation 24.2

$$\frac{d\phi_g}{dz} = \frac{\pi i}{\xi_0}\phi_g + \frac{\pi i}{\xi_g}\phi_0 \exp\left[-2\pi i (sz + \mathbf{g} \cdot \mathbf{R})\right] \qquad [24.3]$$

and

$$\frac{d\phi_0}{dz} = \frac{\pi i}{\xi_0}\phi_0 + \frac{\pi i}{\xi_g}\phi_g \exp\left[+2\pi i (sz + \mathbf{g} \cdot \mathbf{R})\right] \qquad [24.4]$$

Next, we simplify these equations just as we did in Chapter 13 by setting

$$\phi_0(z)_{(sub)} = \phi_0 \exp\left(\frac{-\pi i z}{\xi_0}\right) \qquad [24.5]$$

and

$$\phi_g(z)_{(sub)} = \phi_g \exp\left(2\pi i sz - \frac{\pi i z}{\xi_0}\right) \qquad [24.6]$$

Then the Howie–Whelan equations become

$$\frac{d\phi_{0(sub)}}{dz} = \frac{\pi i}{\xi_g}\phi_{g(sub)} \exp\left(2\pi i \mathbf{g} \cdot \mathbf{R}\right) \qquad [24.7]$$

and

$$\frac{d\phi_{g(sub)}}{dz} = \frac{\pi i}{\xi_g}\phi_{0(sub)} \exp\left(-2\pi i \mathbf{g} \cdot \mathbf{R}\right) + 2\pi i s \phi_{g(sub)} \qquad [24.8]$$

These equations are just as before (equations 13.14 and 13.15) but with the addition of the $2\pi i \mathbf{g} \cdot \mathbf{R}$ term. This additional phase is termed α, hence planar defects are seen when $\alpha \neq 0$

$$\alpha = 2\pi \mathbf{g} \cdot \mathbf{R} \qquad [24.9]$$

These expressions will be particularly useful in two cases:

- When $\mathbf{R}$ = constant.
- Understanding phasor diagrams when defects are present.

We start with a simple stacking fault lying parallel to the surface, as shown in Figure 24.2. In this situation the beams propagate through the upper layer just as if no fault were present. At a depth $z = t_1$, the beams may experience a phase change due to the effect of the translation $\mathbf{R}$, but after that they again propagate as if in a perfect crystal.

In this chapter, we'll see several values of α. A special case occurs when $\alpha = \pm 120°$. This value of α is often encountered since it occurs for fcc SFs. We'll also encounter the case where $\alpha = \pm 180°$; this value arises for some special APBs which are really SFs.

24.3. THE SCATTERING MATRIX

This discussion of the scattering matrix introduces no new concepts. It is just a different way of writing the equations

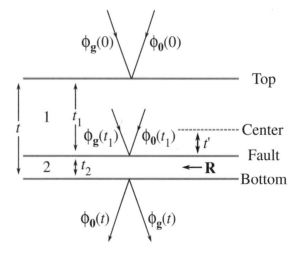

Figure 24.2. A stacking fault lying at depth t_1 in a parallel-sided uniformly thick specimen. The total thickness is t and $t_2 = t - t_1$.

so that, if you are calculating the image contrast, you can program the computer more easily, especially when you have complicated arrays of lattice defects. Our reason for delaying the introduction of the scattering matrix until now is that it is much easier to understand when you can apply it to a specific problem.

In equation 13.20, we showed that in the two-beam case, we can write these simple expressions for ϕ_0 and ϕ_g

$$\phi_0 = C_0 e^{2\pi i \gamma z} \qquad [24.10]$$

and

$$\phi_g = C_g e^{2\pi i \gamma z} \qquad [24.11]$$

Since there are two values for γ, we can express both the **0** and **g** beams as the combination of these two contributions to give

$$\phi_0(z) = C_0^{(1)} \psi^{(1)} \exp\left(2\pi i \gamma^{(1)} z\right) \qquad [24.12]$$
$$+ C_0^{(2)} \psi^{(2)} \exp\left(2\pi i \gamma^{(2)} z\right)$$

$$\phi_g(z) = C_g^{(1)} \psi^{(1)} \exp\left(2\pi i \gamma^{(1)} z\right) \qquad [24.13]$$
$$+ C_g^{(2)} \psi^{(2)} \exp\left(2\pi i \gamma^{(2)} z\right)$$

where the $\psi^{(i)}$ terms tell us the relative contributions of the $\gamma^{(1)}$ and $\gamma^{(2)}$ terms. (We are really saying that both Bloch waves contribute to both the **0** and **g** beams.) We can rewrite equations 24.12 and 24.13 in a matrix form

$$\begin{pmatrix} \phi_0(z) \\ \phi_g(z) \end{pmatrix} = \begin{pmatrix} C_0^{(1)} & C_0^{(2)} \\ C_g^{(1)} & C_g^{(2)} \end{pmatrix} \begin{pmatrix} \exp\left(2\pi i \gamma^{(1)} z\right) & 0 \\ 0 & \exp\left(2\pi i \gamma^{(2)} z\right) \end{pmatrix} \begin{pmatrix} \psi^{(1)} \\ \psi^{(2)} \end{pmatrix} \qquad [24.14]$$

We can express our boundary conditions as

$$C_0^{(1)} \psi^{(1)} + C_0^{(2)} \psi^{(2)} = \phi_0(0) \qquad [24.15]$$

and

$$C_g^{(1)} \psi^{(1)} + C_g^{(2)} \psi^{(2)} = \phi_g(0) \qquad [24.16]$$

which we can now rewrite as

$$\begin{pmatrix} C_0^{(1)} & C_0^{(2)} \\ C_g^{(1)} & C_g^{(2)} \end{pmatrix} \begin{pmatrix} \psi^{(1)} \\ \psi^{(2)} \end{pmatrix} = \begin{pmatrix} \phi_0(0) \\ \phi_g(0) \end{pmatrix} \qquad [24.17]$$

(We actually saw in Section 13.9 that $\phi_0(0)$ is 1 and $\phi_g(0)$ is 0, because $z = 0$ is the top surface.) Now we can use matrix algebra to solve equation 24.17. First rewrite it as

$$C \begin{pmatrix} \psi^{(1)} \\ \psi^{(2)} \end{pmatrix} = \begin{pmatrix} \phi_0(0) \\ \phi_g(0) \end{pmatrix} \qquad [24.18]$$

then rewrite equation 24.18 as

$$\begin{pmatrix} \psi^{(1)} \\ \psi^{(2)} \end{pmatrix} = C^{-1} \begin{pmatrix} \phi_0(0) \\ \phi_g(0) \end{pmatrix} \qquad [24.19]$$

where C^{-1} is just the inverse matrix. Remember that the order is important in matrix multiplication and that $C^{-1}C = I$, the unit matrix.

Therefore we can rewrite equation 24.14 as

$$\begin{pmatrix} \phi_0(z) \\ \phi_g(z) \end{pmatrix} = C \begin{pmatrix} \exp\left(2\pi i \gamma^{(1)} z\right) & 0 \\ 0 & \exp\left(2\pi i \gamma^{(2)} z\right) \end{pmatrix} C^{-1} \begin{pmatrix} \phi_0^{(0)} \\ \phi_g^{(0)} \end{pmatrix} \qquad [24.20]$$

Finally, we can define a new matrix $P(z)$ as the scattering matrix for a slice of thickness z

$$P(z) = C \begin{pmatrix} \exp\left(2\pi i \gamma^{(1)} z\right) & 0 \\ 0 & \exp\left(2\pi i \gamma^{(2)} z\right) \end{pmatrix} C^{-1} = C\,\Gamma\,C^{-1} \qquad [24.21]$$

The matrix $P(z)$ thus gives us the values of the exit wave amplitudes at the bottom of the slice in terms of the incident values. In other words, the matrix $P(z)$ includes all the information to describe the propagation of the beams through the crystal; $P(z)$ is a *propagator* matrix. Notice that z only enters the equation through the Γ matrix in equation 24.21.

24.4. USING THE SCATTERING MATRIX

Now we illustrate the real strength of the scattering matrix approach by considering the effect of a planar fault lying parallel to the foil surface, as we saw in Figure 24.2. The idea is that we now have two slices of material of thickness t_1 and t_2. We can easily calculate $\phi_0(t_1)$ and $\phi_g(t_1)$ using equation 24.20. These values for ϕ_0 and ϕ_g then become the incident values for slice 2. The effect of the translation **R** is to multiply the terms in C_g in the lower slice by a phase factor $\exp(-i\alpha)$, where $\alpha = 2\pi \mathbf{g} \cdot \mathbf{R}$ as usual. The matrix C for slice 2 is then written as

$$C_2 = \begin{pmatrix} C_0^{(1)} & C_0^{(2)} \\ C_g^{(1)} \exp(-i\alpha) & C_g^{(2)} \exp(-i\alpha) \end{pmatrix} \qquad [24.22]$$

We can write down the expression for $\phi_0(t)$ and $\phi_g(t)$ as

$$\begin{pmatrix} \phi_0(t) \\ \phi_g(t) \end{pmatrix} = C_2 P(t_2) C_2^{-1} C_1 P(t_1) C_1 \begin{pmatrix} \phi_0(0) \\ \phi_g(0) \end{pmatrix} \quad [24.23]$$

where the subscripts on C_1 and C_2 just identify the slices. Normally, this equation goes straight into the computer. However, we'll go back and consider a few special points:

- Look at equation 24.23 and set $\mathbf{R} = 0$, so that $C_1 = C_2$. You can see that $P(t) = P(t_1)P(t_2)$. Clearly we could cut the perfect-crystal specimen into many slices and $P(t)$ would always be the product of the scattering matrices for each slice.
- How do we prove equation 24.22? From equation 14.12 we know that a Bloch wave can be written as

$$b(\mathbf{k}) = \sum_g C_g(\mathbf{k}) \exp(2\pi i (\mathbf{k} + \mathbf{g}) \cdot \mathbf{r}) \quad [24.24]$$

If the crystal is displaced by a vector $\mathbf{R}$ then we replace $\mathbf{r}$ with $\mathbf{r} - \mathbf{R}$ (notice the sign, see Section 25.14). (We have just used a "hidden" column approximation.) Equation 24.24 is then written

$$b(\mathbf{k}) = \sum_g C_g(\mathbf{k}) \exp(2\pi i (\mathbf{k} + \mathbf{g}) \cdot (\mathbf{r} - \mathbf{R})) \quad [24.25]$$

$$b(\mathbf{k}) = e^{-2\pi i \mathbf{k} \cdot \mathbf{R}} \sum_g C_g(\mathbf{k}) e^{-2\pi i \mathbf{g} \cdot \mathbf{R}} e^{2\pi i (\mathbf{k} + \mathbf{g}) \cdot \mathbf{r}} \quad [24.26]$$

C_0 is not affected by $\mathbf{R}$ since $2\pi \mathbf{0} \cdot \mathbf{r} = 0$, but C_g is multiplied by $e^{-i\alpha}$.

- If you choose the coordinates appropriately (see Chapter 14) then C is a unitary matrix. In this case, you can find C^{-1} just by reflecting across the diagonal and taking the complex conjugate of each term. This *trick* will allow you to express equation 24.23 explicitly, as given by Hirsch et al. (1977) (omitting a phase factor)

$$\phi_0(t) = [\cos(\pi \Delta k t) - i \cos(\beta) \sin(\pi \Delta k t)] \\ + \tfrac{1}{2}(e^{i\alpha} - 1)\sin^2 \beta \cos(\pi \Delta k t) - \tfrac{1}{2}(e^{i\alpha} - 1)\sin^2 \beta \cos(2\pi \Delta k t') \quad [24.27]$$

$$\phi_g(t) = i \sin \beta \sin(\pi \Delta k t) \\ + \tfrac{1}{2}\sin \beta (1 - e^{(-i\alpha)})[\cos \beta \cos(\pi \Delta k t) - i \sin(\pi \Delta k t)] \\ - \tfrac{1}{2}\sin \beta (1 - e^{(-i\alpha)})[\cos \beta \cos(2\pi \Delta k t') - i \sin(2\pi \Delta k t')] \quad [24.28]$$

In equations 24.27 and 24.28, t' is the distance of the fault below the center of the slice, i.e., we define $t' = t_1 - t/2$ where t_1 lies between 0 and t. (It's a good, but tedious, exercise to derive these equations for yourself.) The right-hand side of equations 24.27 and 24.28 each contains three terms:

- The first term is just what we found in Chapter 13 where the phase factor $\alpha = 0$, i.e., it's just like the perfect crystal.
- The second term is independent of the position of the planar fault because it doesn't depend on t'.
- The third term depends on t' such that both ϕ_0 and ϕ_g change with a periodicity in t' given by Δk^{-1}. So these amplitudes show the same dependence on ξ_g^{eff}. They will both show thickness variations.

You should keep in mind that we derived these equations for a planar defect lying parallel to the parallel surfaces of our specimen and normal to the beam. We can now take these ideas and apply them to planar defects which are inclined to the surface, by calculating the contrast for all values of t' between 0 and t. The important points to remember are:

- The model used in the calculation was a flat interface parallel to the surface of a platelike specimen. You'll see fault fringes when t' varies across the fault, but you don't usually have to consider the fact that either the surface or the fault may be inclined to the beam.
- The concept of the scattering matrix allows you to identify very clearly the effect of the defect on ϕ_0 and ϕ_g.

24.5. STACKING FAULTS IN fcc MATERIALS

We'll begin our discussion of actual examples with the SF in fcc materials. Before we discuss the details of contrast from SFs in fcc materials, we'll summarize the important results which hold for all planar defects:

- The appearance of the image depends on the specimen thickness.
- Pairs of BF/DF SF images are not generally complementary even though we are using a two-beam approximation. Compare to the complementary behavior of the thickness fringes discussed in Chapter 23.

- Planar defects can, in fact, have a thickness. We'll illustrate this concept using overlapping faults in fcc materials (see also Section 6.6).

> Do not assume all faults are the same as in fcc materials!

24.5.A. Why fcc Materials?

There are several reasons for emphasizing the analysis of stacking faults in fcc crystals:

- Many important materials are fcc, including the metals Cu, Ag, Au, and austenitic stainless steel, and the semiconductors Si, Ge, and GaAs.
- Most of the analysis of SFs derives from the study of fcc materials.
- The translations are well known and directly related to the lattice parameter: **R** is either $\frac{1}{6}<\bar{1}\bar{1}2>$ or $\frac{1}{3}<111>$. Notice that these definitions differ by the lattice vector, $\frac{1}{2}<110>$. (Actually there may be small deviations from these ideal values, but we'll ignore them for now.)

We want to learn how to extend this analysis to other fault vectors and avoid making unfounded assumptions when we do extend it. The geometry often encountered is shown in Figure 24.3. If the sample is single crystal, then you need to prepare a specimen with a [111] foil normal, so that you can image long segments of the dislocations lying in the plane of the foil on their (111) glide plane.

You should note that $(11\bar{1})$ is one of three possible planes for an inclined SF. In this case, the translation at the stacking fault will be $\mathbf{R} = \pm \frac{1}{3}[11\bar{1}]$; the phase factor, α, is $2\pi \mathbf{g} \cdot \mathbf{R}$. If you form an image with the $\mathbf{g} = (2\bar{2}0)$ reflection strongly excited, then $\mathbf{g} \cdot \mathbf{R} = 0$ and the fault is out of contrast in both BF and DF. If, instead, you use the reflection $\mathbf{g} = (02\bar{2})$, then $\mathbf{g} \cdot \mathbf{R} = \frac{2}{3}$ or $-\frac{2}{3}$ and $\alpha = \frac{8\pi}{3} = \frac{2\pi}{3} = 120°$ or $-\frac{8\pi}{3} = -\frac{2\pi}{3} = -120°$ (modulo 2π in each case). Notice that if the stacking fault lies parallel to the surface of this (111)-oriented specimen, you must tilt the specimen to see *any* contrast from the SF, i.e., $\mathbf{g} \cdot \mathbf{R} = 0$ for all values of **g** lying in the fault plane.

Figures 24.4A–D show two typical BF/DF pairs of ±**g** strong-beam images from the same SF. In the BF images the outer fringes are the same on both sides of the fault (both gray or both white) while in the DF images one outer fringe is white but the other is gray, as summarized in Figures 24.4E and F. The questions which arise are:

- What determines whether a fringe will be gray or white?
- Why are the two images not complementary?

Note in Figures 24.4E and F that Type A reflections are 200, 222, and 440 while Type B are 111, 220, and 400.

24.5.B. Some Rules

There are some experimental rules:

- Be very careful when you record such a pair of images: record the DP for each image. Be sure to note which of the two bright spots corresponds to the direct beam.
- Use the same strong $hk\ell$ reflection for BF and DF imaging. Therefore, to form the CDF image using a strong $hk\ell$ reflection you must first tilt the specimen so that $\bar{h}\bar{k}\bar{\ell}$ is strong, and then use the beam tilts to move $hk\ell$ onto the optic axis where it will become strong (see Section 22.5). This is confusing, so we recommend that you sacrifice a little image resolution and compare the BF image with a displaced-aperture DF image, rather than a CDF image.

In a modern IVEM, there is almost no loss of resolution between displaced-aperture DF and CDF.

> So to avoid the possibility of confusion about which reflection to use for the DF image (e.g., see Edington 1976, p. 148), just displace the aperture to the strong $hk\ell$ reflection in every case.

This is exactly the opposite of the approach used by Edington, who advocates tilting in $\bar{h}\bar{k}\bar{\ell}$ for the DF image, which reverses the DF contrast in Figures 24.4E and F. Our approach ensures that diffraction from the same $hk\ell$ planes causes the contrast in both BF and DF images.

Then there are some rules for interpreting the contrast:

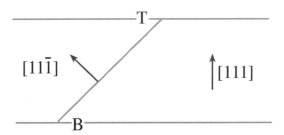

Figure 24.3. A stacking fault in a parallel-sided fcc specimen. The normal to the specimen is [111] and the normal to the SF is $[11\bar{1}]$. T and B indicate the top and bottom of the foil.

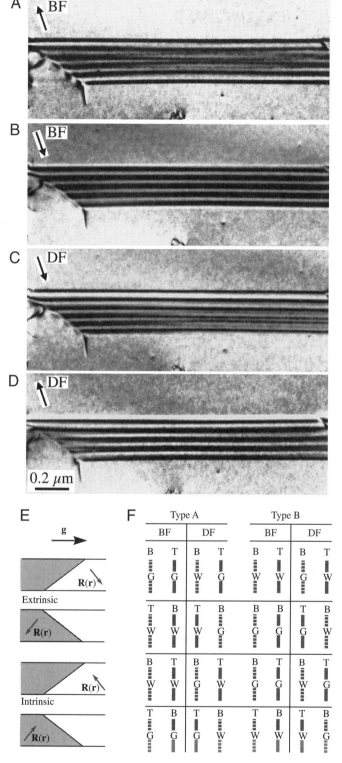

- In the image, as seen on the screen or on a print, the fringe corresponding to the top surface (T) is white in BF if **g·R**>0 and black if **g·R**<0.
- Using the same strong $hk\ell$ reflection for BF and DF imaging, the fringe from the bottom (B) of the fault will be complementary whereas the fringe from the top (T) will be the same in both the BF and DF images.
- The central fringes fade away as the thickness increases. If this seems anomalous, the explanation is in Section 24.10.
- The reason it is important to know the sign of **g** is that you will use this information to determine the sign of **R**.
- For the geometry shown in Figure 24.3, if the origin of the **g**-vector is placed at the center of the SF in the DF image, the vector **g** points away from the bright outer fringe if the fault is extrinsic and toward it if it is intrinsic (200, 222, and 440 reflection); if the reflection is a 400, 111, or 220 the reverse is the case.
- Don't forget that, as we said at the start of Chapter 22, any contrast must be >~5–10% to be visible to the eye, so any intensity change due to **g·R** effects is only detectable if **g·R**>0.02 (unless you can digitize and process your analog TEM image). With experience, you'll find there is an optimum thickness to view defect contrast, before absorption effects make it difficult. You must also carefully select **s** so the background intensity in the matrix around the defect is gray and this maximizes visibility of lighter and darker fringes.

> So just because you can't see a defect doesn't mean it isn't there, or that **g·R** = 0.

As we said, these complex rules are summarized in Figures 24.4E and F. Although they are very useful, in practice you should remember that they were derived for a very special combination of **R** and **g** in fcc materials. Some important examples of **g·R** are given in Table 24.2. As we'll describe in Section 24.11, you should use a computer program to check the contrast.

24.5.C. Intensity Calculations

Now let's consider intensity calculations using the column approximation, which we briefly discussed in Section 13.11. If the fault cuts the column at a depth t_1, we can deduce from equations 24.22 and 24.23 that

Figure 24.4. (A–D) Four strong-beam images of an SF recorded using ±**g** BF and ±**g** DF. The beam was nearly normal to the surfaces; the SF fringe intensity is similar at the top surface but complementary at the bottom surface. The rules are summarized in (E) and (F) where G and W indicate that the first fringe is gray or white, and (T,B) indicates top/bottom.

Table 24.2. Values of g·R for Some Common R/g Combinations

	R	g	$\alpha = 2\pi \mathbf{g \cdot R}$ (mod 2π)
SF in fcc	$\tfrac{1}{3}[111]$	(111), (220), (113)	$2\pi/3$
SF in fcc	$\tfrac{1}{3}[111]$	(113)	$4\pi/3$
Translation at APB in Fe$_3$Al	$\tfrac{1}{2}[110]$	(100)	π
Small **R**, e.g., NiO	any	**g** or **s** or ξ_g differ slightly	δ

$$\phi_g = \frac{i\pi}{\xi_g}\left\{\int_0^{t_1} e^{-2\pi i s z}dz + e^{-i\alpha}\int_{t_1}^{t} e^{-2\pi i s z}dz\right\} \quad [24.29]$$

which gives

$$\phi_g = \frac{i\pi}{s\xi_g} e^{-2\pi i s t_1}\left\{\sin(\pi s t_1) + e^{-i\alpha}\sin(\pi s(t-t_1))\right\} \quad [24.30]$$

We rearrange equation 24.30 to give an expression for the intensity, $I_g (= \phi_g \phi_g^*)$. This rearrangement involves a little manipulation

$$I_g = \frac{1}{(s\xi_g)^2}\left\{\sin^2\left(\pi s t_1 + \frac{\alpha}{2}\right) + \sin^2\left(\frac{\alpha}{2}\right) \right.$$

$$\left. - \sin\left(\frac{\alpha}{2}\right)\sin\left(\pi s t + \frac{\alpha}{2}\right)\cos(2\pi s t')\right\} \quad [24.31]$$

where $t' = t_1 - t/2$ as before. So the contrast depends on both the thickness and the depth. Note that $t/2$ is the center of the foil. Since α is fixed for a particular defect, let's fix t. Then equation 24.31 becomes

$$I_g \propto \frac{1}{s^2}\left\{A - B\cos(2\pi s t')\right\} \quad [24.32]$$

Now we have cosine depth fringes or defect thickness fringes, just as we did for the perfect crystal.

- The thickness periodicity depends on s^{-1}.
- The intensity varies as s^{-2}.

We could have derived this equation from equation 24.28 with more work. However, the value of the scattering matrix approach is that we don't derive the analytical expression but just run the computer.

In Chapter 26 we will discuss this SF contrast in terms of phasor diagrams, which give a graphical way to represent these equations.

24.5.D. Overlapping Faults

It is interesting to extend this analysis to the case of overlapping faults. Taking the analytical approach, we can extend equation 24.29 to the case of two overlapping faults, the first at depth t_1 and the second at depth t_2

$$\phi_g = \frac{i\pi}{\xi_g}\left\{\int_0^{t_1} e^{-2\pi i s z}dz + e^{-i\alpha}\int_{t_1}^{t_1+t_2} e^{-2\pi i s z}dz \right. \quad [24.33]$$

$$\left. + e^{-i(\alpha_1+\alpha_2)}\int_{t_1+t_2}^{t} e^{-2\pi i s z}dz\right\} \quad [24.33]$$

An experimental illustration of a somewhat more complex situation, involving several overlapping SFs, is shown in

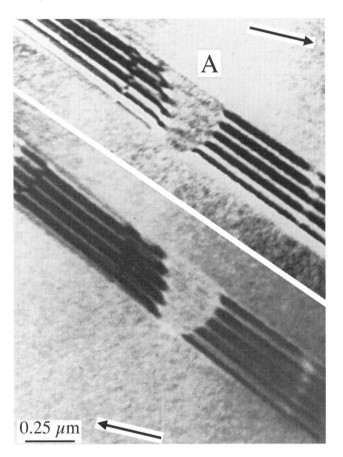

Figure 24.5. Two ±**g** BF images of overlapping SFs in fcc steel, with the direction of **g** indicated. The faults are very close together. When three faults overlap the effective value of **R** is 0 so the contrast disappears.

Figure 24.5. We can see two very interesting features in this image:

- The BF/DF contrast is not complementary.
- It sometimes appears that there is no contrast, even when we know that there are overlapping SFs. This can happen if, e.g., three SFs overlap on adjacent (or nearly adjacent) planes; then the effective **R** can be $3 \times \frac{1}{3}[11\bar{1}]$, which is a perfect lattice vector and can therefore appear to give $2\pi\mathbf{g}\cdot\mathbf{R} = 0$.

We will return to this topic in Section 26.6 where we'll show that some planar defects, such as the extrinsic SF in Si or the dissociated {112} twin boundary in some fcc metals, really have a thickness. We can then analyze the contrast from such interfaces using the overlapping-fault model.

24.6. OTHER TRANSLATIONS: π AND δ FRINGES

We discussed the $L1_0$ structure of NiAl in Section 16.5. This intermetallic is an example of a large group of materials which can contain a different type of RB. If the Ni atoms sit at the corners of the cell in one crystal region but the Al atoms sit at the corners in another part of the crystal, then the two crystal regions are related by a translation of $\frac{1}{2}[111]$. The two crystals would otherwise still be perfectly aligned but are separated by this RB, which we call an APB.

Similarly in the $L1_2$ structure of the intermetallic Ni_3Al, we could have the Al atoms on the corners of the unit cell in one part of the crystal and displaced by $\mathbf{R} = \frac{1}{2}[110]$ in the adjacent region. (We can actually form six nonequivalent APBs in this structure.) The crystal structure looks like fcc but the Al atoms are at the corners of the unit cell (forming the simple-cubic superlattice) with the Ni at the face-centered positions. The easy way to appreciate this RB is to think what would happen if the alloy were completely disordered: there would be no planar defect. This RB can be imaged using the (100) reflection. Notice again that for a disordered structure, the {100} reflections would be absent if the alloy were disordered; the {100} planes are said to give rise to superlattice reflections; these reflections would be forbidden if the material were disordered. For this case we can readily show that the phase factor $\alpha = \pi$, so the fringes we see are called π fringes. The structure of this interface is shown schematically in Figure

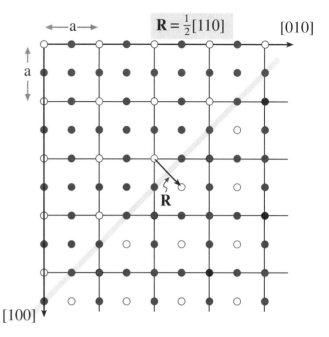

Figure 24.6. Schematic of an interface in the intermetallic Ni_3Al showing how the two structures link coherently. The phase factor at such an interface is π and the fringes seen in the image are called π fringes.

24.6. These π fringes can give symmetric fringes in DF and BF and complementary BF/DF pairs.

Similar RBs are very common in oxides because the unit cell is often quite large, giving more opportunities to form such interfaces. The interface shown in Figures 24.7A and B has been called both an SF and an APB in the spinel. These interfaces can show all the features we discussed in Section 24.5 for SFs in fcc materials, and those we've just discussed depend on which reflection you use. You can again see a change in contrast in Figure 24.7C when APBs in TiO_2 overlap as shown schematically in Figure 24.7D (Amelinckx and Van Landuyt, 1978). In Figures 24.7A,B, if you image the fault using the 220 reflection $2\pi\mathbf{g}\cdot\mathbf{R} = \pi$ and so you'll see SF fringes. If however, you image using 440, $2\pi\mathbf{g}\cdot\mathbf{R} = 2\pi$, so you'll only see residual contrast (because **R** is not exactly $\frac{1}{4}[101]$).

The APB shown in Figure 24.8 is different yet again. This planar defect in GaAs is also known as an IDB (Section 24.1). The fringes you see are caused by a translation, but **R** is not related in a simple way to the structure of the crystal (Rasmussen *et al.* 1991). As shown in the diagram, the translation is present because there is a small relaxation of the Ga-Ga and As-As bonds at this {110} interface. The value of **R** was determined to be 0.19Å with a statistical uncertainty of ±0.03Å. These fringes are then known as δ fringes (because they are only small transla-

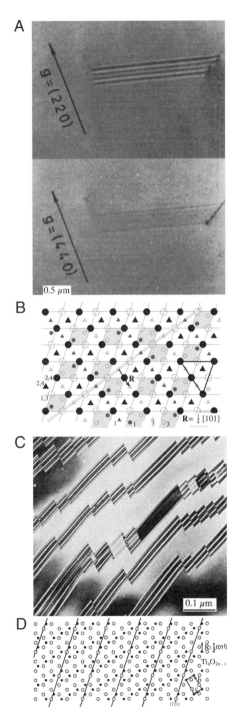
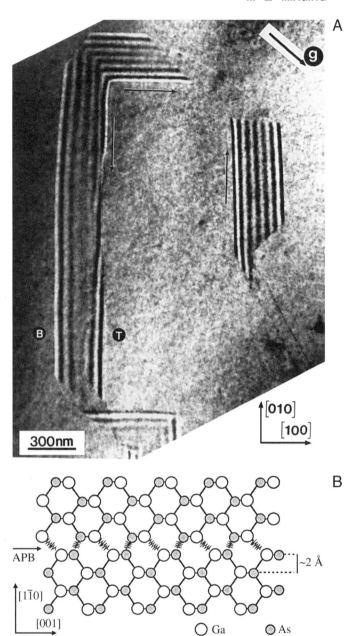

Figure 24.7. (A) Pair of BF images and (B) schematic of an SF in spinel; this interface is also known as an APB since the SF translation vector is a perfect (sub)lattice vector in the underlying fcc oxygen sublattice as shown in (B). Large circles are O anions at different heights (1,3) and (2,4); the small circles and triangles are cations at different heights (1–4 as indicated). (C) APBs can overlap just as SFs can as shown by these faceted APBs in TiO_2. Many of the facets give quite similar contrast but those near the center are strikingly different because of overlap. The schematic (D) shows a series of APBs, each of which is formed by a translation which has little effect on the oxygen sublattice.

Figure 24.8. (A) A faceted APB (or IDB) in GaAs with (B) a schematic of the $(1\bar{1}0)$ facet. The translation is caused by the difference in length of the Ga-Ga and As-As bonds and does not correspond to any length in the GaAs lattice.

tions and are not related in a simple way to 2π). The use of image-simulation programs, which are necessary to determine **R** (remember that the wavelength of the 200-kV electrons used for the measurement is itself 0.025Å), is discussed in Section 24.13.

24 ■ PLANAR DEFECTS

Table 24.3. Examples of Special Phase Boundaries

Boundary	Example of material	Features
Ferromagnetic domain boundaries	NiO	
Ferroelectric/piezoelectric boundaries	BaTiO$_3$	Small tetragonal distortion
Composition boundary	GaAs/AlGaAs	ξ_g is different on two sides of boundary, even for perfect lattice matching
Structure boundaries	α-SiC/β-SiC	
	hcp-Co/fcc-Co	
Composition/structure	Nb/Al$_2$O$_3$	
	Al/Cu	
	α-Fe/Fe$_3$C	

24.7. PHASE BOUNDARIES

We list a few special phase boundaries in Table 24.3.

An example of a PB is shown in Figure 24.9. In NiO, which is ferromagnetic, some of the planes rotate when the structure changes from cubic symmetry below the Curie temperature. Now we can also define the cubic structure as rhombohedral with α = 60° in the rhombohedron. Below the Curie temperature, the rhombohedral angle is distorted by only 4.2′ from the true 60°. Therefore, most **g**-vectors will rotate through a very small angle and hence produce a change in the value of **s**. However, as you can see in Figure 24.9, this small rotation can readily be detected by the change in contrast and the faint fringes at the phase boundary.

We can have overlapping PBs, so the warning is the same: be very wary and use tilting experiments.

Figure 24.9. The ferromagnetic material NiO undergoes a structural change from cubic to distorted rhombohedral at the Curie temperature. Although the distortion in the rhombohedral structure is very small, it causes a detectable rotation of the lattice planes which results in the δ fringes in the image.

24.8. ROTATION BOUNDARIES

What can we learn about rotation boundaries when the rotation angle is greater than about 0.1°? Unfortunately, the answer is "not a lot," unless we have defects which accommodate the rotation. Then we are into the subject of diffraction contrast of line defects in interfaces. However, with care you may be able to excite **g** in one grain or in both by tilting the specimen. The difficulty, of course, is that s_g is likely to be different in each material. Complications will also arise if other defects are present, since you may or may not see those defects. Examples of such interfaces are shown in Figure 24.10.

24.9. DIFFRACTION PATTERNS AND DISPERSION SURFACES

You read in Chapter 17 that what you see in an image must be related to what happens in the DP, which in turn is determined by how the Ewald sphere intersects the reciprocal lattice. Figure 17.5 showed that a planar defect which is inclined to the surface of a parallel-sided specimen will give rise to relrods. Therefore, a planar defect in a parallel-sided specimen will produce at least two spots in the diffraction pattern. Since most specimens are wedges (see Figure 17.4), and the planar defect will, in general, be inclined to both surfaces, the relrod geometry is even more complex. Figure 24.11 shows lines normal to each interface and their associated relrods. You can appreciate that when the Ewald sphere cuts these relrods, several spots may appear in the diffraction pattern. Now we need to relate these relrods to the fringes we see in the image. This model would predict that we would not produce fringes when **s** = 0, so we should modify what we did in Figure 17.15. The periodicities of the fringes in the image are inversely related to the distances (M$_1$N, and M$_2$N) between the spots in the DP.

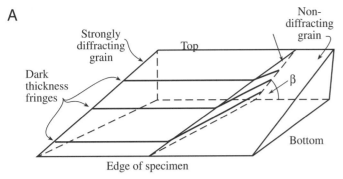

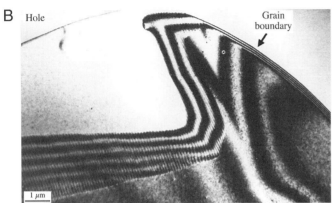

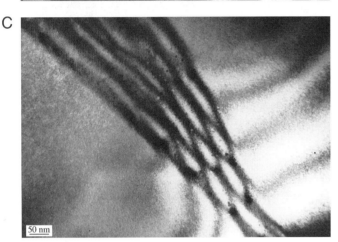

Figure 24.10. (A) If the adjoining grains are rotated so that they do not share a common reflection, images can be formed where only one of the grains diffracts. As shown in (B), the thickness fringes associated with the wedge-shaped foil merge into the thickness fringes associated with the inclined interface. (C) If the foil is tilted so that the same (though not coincident) reflection is excited in both grains, the number of fringes in the interface increases with each incremental increase in the wedge thickness.

> At $s = 0$, the fringe spacing is determined by $\Delta k = \xi_g^{-1}$; the spacing of the fringes is ξ_g.

When a planar defect is present in the specimen, the two branches of the dispersion surface are not only coupled

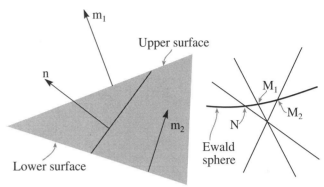

Figure 24.11. In a wedge specimen, the planar defect will, in general, be inclined to both surfaces and the relrod geometry is complex. The fringe spacing in the image is related to the reciprocal of the distances M_1N and M_2N.

along a tie line normal to the surface of the specimen but also along that normal to the planar defect. However, when $s = 0$, the thickness periodicity in the image corresponds to the extinction distance. When we relate this to the region G in the reciprocal lattice, the two relrods (which are a kinematical construction) must actually separate to give the two hyperbolas shown in Figure 24.12, which is why we drew Figures 17.15 and 23.6 as we did.

> If you see fringes in the image, spots will be present in the DP.

The spots in the DP are associated with points M and N in Figures 24.11 and 24.12.

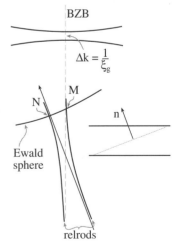

Figure 24.12. The dispersion-surface construction for an inclined planar defect in a parallel-sided specimen. (Compare with Figures 17.15 and 23.6.) For simplicity, we show the hyperbolas due to the defect alone, not the extra effects that would arise in a wedge specimen.

24.10. BLOCH WAVES AND BF/DF IMAGE PAIRS

In Chapter 14 we saw that, in a crystal, the electron must propagate as Bloch waves, and yet we have not mentioned Bloch waves in our discussion of thickness and bending so far. Most of the analysis of this topic is beyond the scope of this text, but it is important to understand the basic ideas, particularly since they will also apply to scattering from defects in the crystals. Remember that ξ_g is a direct consequence of having two Bloch waves. The important message here is: don't let the words overawe you.

The idea is quite simple. Since we have two Bragg beams excited, then we must have two Bloch waves in the crystal. The propagation vectors of these two waves are $\mathbf{k}_1$ and $\mathbf{k}_2$, with the difference $|\Delta \mathbf{k}|$ being given by s_{eff}. We see a thickness dependence in the image because the two waves are interfering. The only two waves which are really present in the crystal are the two Bloch waves. It's the beating of these two waves which gives rise to thickness effects.

In the two-beam case, the Bloch waves, 1 and 2, are channeled along and between the atom columns (see Figure 14.2). A fault may change the channeled wave into the

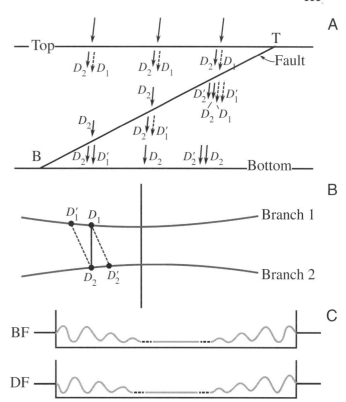

Figure 24.14. The absorption of the branch-1 Bloch wave near the top surface, T, of the specimen and its re-creation when the planar defect is near the bottom, B, determines the contrast we see. (A) shows which Bloch waves are present at the different depths in the specimen, (B) shows how the Bloch waves are coupled along tie lines joining the two branches of the dispersion surface, and (C) shows the resulting contrast.

nonchanneled one, as you can see in Figure 24.13. The effect of the planar defect is simply to couple the Bloch waves; in other words, the defect links the different branches of the dispersion surface. The noncomplementary contrast at SFs in fcc metals is directly explained by this coupling.

As soon as the beam enters the specimen we excite Bloch waves 1 and 2. Therefore, in Figure 24.14A, the two Bloch waves 1 and 2 are shown everywhere at the top surface of the foil. The planar defect links points D_1 and D_2 on the two branches of the dispersion surface, as shown in Figure 24.14A, along the tie line, D_1D_2' and $D_1'D_2$. We'll analyze the three situations shown in Figure 24.14B, which correspond to the planar defect being close to the top, the middle, and the bottom of the specimen. The key feature is that, as we saw in Section 14.6, Bloch wave 1, which has the larger $\mathbf{k}$-vector, will be preferentially absorbed. It is actually totally absorbed in thicker specimens.

- When the planar defect is close to the top surface (as occurs near T), waves 1 and 2 are both coupled (or scattered) to the other branch of

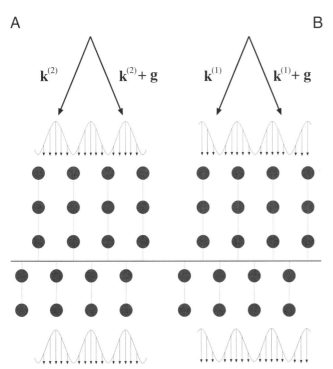

Figure 24.13. Bloch waves 1 and 2 are channeled along and between the atom columns, respectively, until they meet the fault. There the atomic columns are translated so that the channeled Bloch wave may become the nonchanneled one.

the dispersion surface so we form four Bloch waves (but with only two **k**-vectors). Both Bloch waves which are associated with the upper branch of the dispersion surface (wave vector $\mathbf{k}_1$) are preferentially absorbed, but the waves D_2 and D_2' both reach the lower surface. There, they interfere to give the thickness fringes even though they are both associated with the lower branch of the dispersion surface; D_2' retains a "memory" of D_1.

- When the fault is close to the middle of the specimen, the branch-1 Bloch wave is absorbed before it reaches the planar defect but a new Bloch wave D_1' is formed at the defect. However, while traversing the other half of the foil, this wave is also absorbed so that only wave D_2 reaches the lower surface. Thus the electrons can propagate through the specimen (we can see through it) but there are no thickness fringes, because only one Bloch wave survives. However, we can still image defects in these thicker areas, as you'll see if you look back at Figure 23.10.
- At the lower surface, B, only wave D_2 survives to reach the planar defect but it now produces a new wave D_1', which can reach the lower surface, recombine with Bloch wave D_2, and produce thickness fringes. The resulting contrast is summarized in Figure 24.14C.

Bloch-wave absorption is a critical factor in explaining the appearance of contrast from planar defects. The part of this argument which is not intuitive is the fact that D_2' retains a memory of D_1; this memory allows it to interfere with D_2 to produce the thickness fringes near the top of the specimen, even though no Bloch wave from branch 1 reaches the bottom of the specimen. We'll refer you to the article by Hashimoto *et al.* (1962) for further discussion on this topic.

24.11. COMPUTER MODELING

From the discussion in Sections 24.5 and 24.6, you will realize that α and π fringes are usually understandable as long as you know what the defect is, and as long as it's not actually a set of overlapping defects. The contrast from δ fringes is much more complex and combinations of α, π, and δ fringes are difficult! The situation will become even more complicated if you want to understand the contrast occurring when other defects interact with these planar faults. A computer program is then really the only way to analyze the contrast from these defects.

The first program to attempt the task of simulating two-dimensional images of planar defects is described in the book by Head *et al.* (1973) (see Section 1.5). One of several modern approaches is Comis, a Unix program, which is also available in a Macintosh version. We will mention some of the features of these programs to help you select one, but leave the detailed descriptions to the appropriate manuals. The most important reason for using any program must be your desire to understand the contrast and thus characterize the defect.

> These programs are tools to assist you toward the goal of quantitative analysis of diffraction contrast, but you always need the fully quantitative experimental image, too.

You need an accurate simulation of the image you see in the microscope. The fact that the image varies with depth, thickness, **g**-vector, etc., is actually to your advantage, since you then have many variables, all of which you must be able to measure to achieve a good match with your experimental image. From the point of view of quantitative analysis, a one-dimensional line (intensity) profile is as valid as a two-dimensional image. Of course, if you can compare the contrast in an image and a simulation point by point, then you can have much greater confidence in the matching. The two-dimensional simulated image is also more viewer-friendly!

A great advantage of more powerful computers is that you can also test the effect of specimen geometry more readily. Thus, for example, Viguier *et al.* (1994) have shown using the Cufour simulation package (Schäublin and Stadelmann 1993) that the rules for fringe contrast given by Gevers *et al.* (1963) will not work if the specimen is tilted such that it intersects the bottom of the foil above the point where it intersects the top! You can understand this situation more easily by looking at the specimen geometry shown in Figure 24.15.

Image simulation tells us that **g·R** must be >0.02 to produce visible fringes, and you don't need to know the local structure at the planar defect when determining this condition. You could, in principle, detect smaller values of **R** by using larger **g**-vectors, but in practice it then becomes more difficult to set up a well-defined diffraction geometry.

The next two sections are rather specialized and you may wish to leave them until much later, especially if you don't have access to a suitable program, or until you are prepared to write your own. Do consult the key references and list of available programs in Section 1.5 before writing your own program. The subject is just as relevant to the topics of Chapters 25 and 26, but we include it in this

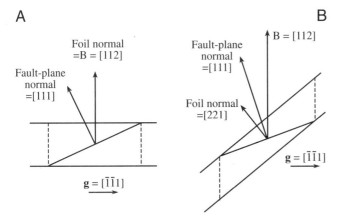

Figure 24.15. Many of the "rules" for predicting the contrast from planar "defects make certain assumptions about the geometry of the defect relative to the surface of the specimen which may not always hold. Here it is demonstrated that the intersection of the planar defect with the upper specimen surface may be lower than the intersection with the bottom surface. This geometry can cause a reversal in the rules.

chapter mainly because the analysis of planar defects is the most straightforward application.

24.12. THE GENERALIZED CROSS SECTION

Head *et al.* (1973) presented a method and a computer program for the computation of BF and DF images of line and planar defects. The source code is given in their book and is available from the WWW. You should note several important features of this program:

- It uses the two-beam theory of electron diffraction.
- It uses the column approximation.
- The simulated image can be displayed as a halftone image rather than as intensity profiles.

This program was so successful, in part, because Head *et al.* were able to calculate the images quite quickly in spite of the fact that the computers available to microscopists were often not particularly powerful in ~1970. The calculations used a concept which they called the *generalized cross section* (GCS). (The GCS is not a scattering cross section, it is actually a slice through the specimen.) The GCS can be used when the displacement field, u_k, satisfies the requirement that

$$u_k(x, y, z) = u_k(x, 0, z + cy) \quad [24.34]$$

Here, c is a constant and the foil is imagined to be laterally infinite. When this requirement is satisfied, the calculation of u_k is greatly simplified. One such situation is the important case where several dislocations and their associated fault planes are all parallel to one another. Then you only need to calculate the many-beam Howie–Whelan equations on the plane $y = 0$. The displacement field for two columns y_1 and y_2 will only differ by a translation along the column, i.e., the z-direction. You don't want to repeat calculations you've already done; just calculate the image on a mesh in the x-y plane.

Examples of comparisons between experimental and simulated images using Comis are shown in Figure 24.16. This package is particularly attractive since it also performs elasticity calculations for simple defect configurations. You'll also find it instructive for simulating the effect of changing different parameters, such as:

- Change the accelerating voltage to see the extinction distances change.
- Change the absorption parameters to see the loss of SF-fringe contrast near the middle of the foil ($z = 0.5t$).

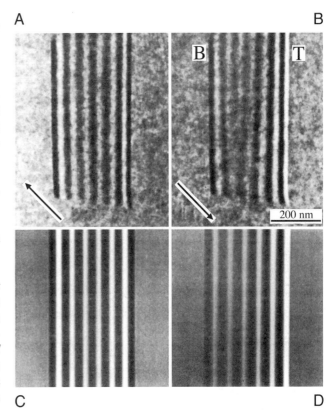

Figure 24.16. (A) Experimental BF image of an APB with $\mathbf{g} = 220$. (B) DF image of the same defect, $\mathbf{g} = \bar{2}20$. (C,D) Corresponding simulated images.

- Change the number of beams contributing to the image; how good is your two-beam assumption?
- Look at how reversing **g** changes the geometry of the image.
- Compare BF and DF as you vary the value of s_g.

24.13. QUANTITATIVE IMAGING

One of the important applications of diffraction-contrast images is the detailed characterization of defects. With the improvements in the TEM, particularly in resolution and drift, we are now able to pay more attention to the fine structure of defects and this requires *quantitative* image analysis. In particular, we need to use the actual intensity levels in the image. One obstacle for quantitative analyses has been the uncertainty in the background-level intensity caused by inelastic scattering. As energy-filtered images become more widely available (see Chapter 40) this problem will disappear. Direct digital recording of the intensities, using a CCD camera, will also make a quantitative analysis more tractable, eliminating uncertainties associated with the calibration of the response of the emulsion of the photographic film. (See also Section 30.4.)

With these new applications of diffraction-contrast images in mind, improved simulation programs have become essential. An ideal program will be versatile, but user-friendly; it will allow you not only to calculate the image but also give defect-interaction geometries. In simulating images of crystal defects you'll encounter several problems which are almost independent of one another. You must:

- Define the geometry of the defects and of the specimen (the diffracting conditions).
- Calculate the displacement field associated with the defects.
- Propagate and scatter the electron beams throughout the foil (i.e., solve the Howie–Whelan equations).

We've already discussed the theoretical basis of the diffraction process, so we'll now illustrate some of the numerical methods which you can employ for different defect geometries. In Chapter 25 we'll consider other types of defects which may be analyzed, methods for defining them, and how you can calculate the displacement field.

24.13.A. Theoretical Basis and Parameters

We'll use the Comis program as an example of the considerations which go into a simulation program. One message which you should certainly understand from this discussion is that you must be very cautious when using any program to simulate images. All such programs make assumptions and simplifications.

> As always, when using the computer to simulate TEM images: beware of the black box. Don't automatically believe everything that comes out of it.

Comis is based on the Howie–Whelan dynamical theory of electron diffraction and therefore neglects diffuse scattering. We'll follow the approach given by Howie and Basinski (1968).

- Use the deformable-ion approximation to describe how the crystal is influenced by the displacement field, **R**. In this model, the potential at **r** in the deformed crystal is assumed to be equal to the potential at the point $\mathbf{r} - \mathbf{R}(\mathbf{r})$ in the perfect crystal. The model is good unless $\mathbf{R}(\mathbf{r})$ varies too rapidly.
- Extend the Howie–Whelan approach to many beams and avoid the column approximation.

The resulting equations are basically the same as those we derived in Chapter 13, so don't be put off by their appearance.

> We are now including terms which allow for a variation in *x* and *y*: these terms were specifically excluded in Chapter 13 when we made the column approximation.

The equations are now written as

$$\frac{\partial \phi_g(r)}{\partial z} =$$

$$i\pi \sum_h \left(\frac{1}{\xi_{g-h}} + \frac{i}{\xi'_{g-h}} \right) \phi_h e^{2\pi i \left((s_h + s_g)z - (\mathbf{h}-\mathbf{g}) \cdot \mathbf{R} \right)} \quad [24.35]$$

$$\theta_x \frac{\partial \phi_g}{\partial x} - \theta_y \frac{\partial \phi_g}{\partial y} + \frac{i}{4\pi \chi_z} \left(\frac{\partial^2 \phi_g}{\partial x^2} + \frac{\partial^2 \phi_g}{\partial y^2} \right)$$

As usual, χ is the incident-beam wave vector in vacuum, **g** is a particular diffraction vector, and **h** represents all the other possible diffraction vectors; you might like to compare this equation with equation 13.8. We have defined two new parameters to take account of the direction of the beam

$$\theta_x = \frac{(\chi + \mathbf{g})_x}{\chi_z} \quad \text{and} \quad \theta_y = \frac{(\chi + \mathbf{g})_y}{\chi_z} \qquad [24.36]$$

The x-y plane in the reciprocal lattice contains the dominant reflections, and z is almost parallel to the incident beam. The number of beams you can include in the calculation is limited only by the capacity of your computer. The standard default for a program would be to select only beams on the systematic row. However, nonsystematic beams can substantially influence your image, so it is useful if you can include sets of beams which are coplanar with the systematic row. We define the deviation of the crystal orientation from the exact Bragg condition by specifying the wave-vector components χ_x and χ_y; the latter applies when reflections outside the systematic row are included. Now we can calculate *all* the deviation parameters, $\mathbf{s_h}$; there are many beams and each $\mathbf{s}$ can be different.

Each extinction distance ξ_g is defined as the ratio $\chi_g/|U_g|$, as usual with U_g being a Fourier component of the perfect-crystal potential. The Fourier components can be calculated from X-ray scattering factors, using the Mott expression (Mott and Massey 1965). For most situations the scattering angle is small enough, so the X-ray scattering factors may, in turn, be calculated using the nine-parameter Gaussian fit given by Doyle and Turner (1968).

You'll need to know the unit cell for your material and the Debye–Waller factors. Comis, for example, can then automatically calculate ξ_g using a built-in table of the Doyle–Turner parameters. The Debye–Waller factor (B) is related to the (mean-square) vibrational amplitude of an atom on a lattice site. So it is a temperature-sensitive term. You need to know B if you want to convert X-ray structure factors to electron structure factors (or vice versa) at a given temperature, or if you want to compare structure factor measurements taken at different temperatures. When you calculate extinction distances, the Debye–Waller factors are essential to determine the effect of temperature.

Equation 24.35 is only valid if your crystal has a center of symmetry, otherwise you have to redefine ξ_g. Simulations involving noncentrosymmetric crystals can be performed, but you have to replace the ξ_g and ξ_g' with complex quantities and then defining all the parameters is really difficult (Gevers *et al.* 1966) (see also equation 14.2).

You can take account of absorption effects in the usual way by adding imaginary Fourier components, U_g' (Yoshioka 1957); the absorption distances, ξ_g', are then defined as we discussed in Section 23.8. There are equations which a program can use to estimate the absorption distances, e.g., the linear relation $|U_g'|/|U_g| = a + b|\mathbf{g}|$, as suggested by Humphreys and Hirsch (1968), or you could specify each individual absorption distance directly.

24.13.B. Apparent Extinction Distance

The program developed by Head *et al.* was based on the two-beam approximation. The success of such calculations relies on the fact that ξ_g may be replaced by an apparent extinction distance, $\xi_g^a < \xi_g$. This substitution compensates for scattering into beams that are not included in the two-beam calculation. The term ξ_g^a depends on the t and must be estimated in each individual situation, e.g., by fitting the simulated image to your experimental image. For a quantitative image analysis it is important that you should have as few adjustable parameters as possible; using the many-beam program eliminates the need to use the parameter ξ_g^a. Alternatively, you may determine ξ_g^a by comparing simulated thickness fringes calculated using many-beam and two-beam approximations.

24.13.C. Avoiding the Column Approximation

You can perform simulations with or without the column approximation. With the column approximation, you only keep the first term on the right-hand side of equation 24.35. The equations are reduced to a system of ordinary differential equations which the program must solve at each image point (x, y). In practice, the equation is solved on the nodes of a mesh (columns) using a fifth-order Runge–Kutta integration routine (which you, or the program, can look up when you need it). You need to choose the size and "resolution" of the mesh. As we'll see in Chapter 26, there are situations where the column approximation will not be acceptable (Howie and Sworn 1970).

Without the column approximation, equation 24.35 gives us a system of coupled partial differential equations. The boundary conditions (at $z = 0$) can be generally written in the form

$$a_g \phi_g + b_g \frac{\partial \phi_g}{\partial x} = c_g \qquad [24.37]$$

where we're ignoring changes in the y-dimension.

You can use fixed boundary conditions (Howie and Basinski 1968):

- The foil is divided into thin slices of thickness Δz. You should not confuse this with the multi-slice method for lattice-image simulation which we'll see in Chapter 29; we are still using

Howie–Whelan equations. Then, equation 24.35 is integrated, using the column approximation, through the first slice, i.e., from $z = 0$ to $z = \Delta z$, at all the mesh points.
- The corrections to the column approximation, i.e., the terms containing derivatives with respect to x and y, are then evaluated by interpolation and included.
- The procedure is repeated until the exit surface of the foil is reached.

With this procedure, you are actually applying the column approximation to the outer boundary of the mesh. So, in equation 24.37, $a_g = 1$, $b_g = 0$, and $c_g = \phi_g$ at the initial surface. In order to avoid distortion of the image, you must choose the step size, Δz, carefully and be sure that the distance between columns (mesh size) is small enough (Anstis and Cockayne 1979).

24.13.D. The User Interface

You'll want to run your program interactively so it should include commands which allow you to change parameters easily. Ideally, it will allow you to access each command through the keyboard using a menu. In Comis, for example, certain standard menus are available for special purposes. The user can also build (and save) menus interactively. This allows all the relevant parameters and commands for a particular problem to be present within a single menu. At any time, all the commands are available through the keyboard.

Although typical simulations may be performed in a matter of seconds, many-beam calculations including several dislocations may require more CPU time. For this situation, Comis includes a "submit" command which will start a batch job based on your current data and parameters. Thus, the interactive mode may be used as a convenient way of submitting several jobs with varying parameter values.

For many problems, a purely visual comparison of experimental and simulated images is sufficient to allow you to interpret your image. In these situations you can often find a ξ_g^a such that the simulations can be carried out with only two beams (Head et al. 1973). However, since many parameters are involved in the image-matching process, it is best to eliminate as many unknown variables as possible. Many-beam calculations are even more important for quantitative analyses.

CHAPTER SUMMARY

The key points discussed in this chapter are:

- We see contrast from planar defects because the translation, **R**, causes a phase shift $\alpha = 2\pi \mathbf{g} \cdot \mathbf{R}$.
- In the two-beam case, we can derive analytical expressions to describe the contrast.
- We can use the scattering-matrix method in the two-beam case and can readily extend it to more complicated multibeam situations.

Many different types of planar defect can be studied. You should be careful not to assume that all defects behave the same as SFs in fcc materials.

There is a direct relationship between the information in the images and that in the DPs, which you can understand using the concept of the relrod.

You need to understand how Bloch waves behave to explain why BF/DF pairs of images are not complementary, and why the contrast from planar defects can disappear in the "middle" of the image. The latter is a result of preferential absorption of certain Bloch waves.

You can now use computer modeling of diffraction-contrast images of planar defects to perform quantitative analysis and image matching.

REFERENCES

General References

Amelinckx, S. and Van Landuyt, J. (1978) in *Diffraction and Imaging Techniques in Material Science,* **1** and **2** (Eds. S. Amelinckx, R. Gevers, and J. Van Landuyt), 2nd edition, p. 107, North-Holland, New York.

Christian, J.W. (1975) *The Theory of Transformations in Metals and Alloys,* Part 1, 2nd edition, Pergamon Press, New York.

Edington, J.W. (1976) *Practical Electron Microscopy in Materials Science,* Van Nostrand Reinhold, New York.

Forwood, C.T. and Clarebrough, L.M. (1991) *Electron Microscopy of In-*

terfaces in Metals and Alloys, Adam Hilger, New York. Invaluable for anyone studying interfaces by TEM.

Head, A.K., Humble, P., Clarebrough, L.M., Morton, A.J., and Forwood, C.T. (1973) *Computed Electron Micrographs and Defect Identification,* North-Holland, New York.

Rasmussen, D.R. and Carter, C.B. (1991) *J. Electron Microsc. Techniques* **18**, 429.

Sutton, A.P. and Balluffi, R.W. (1995) *Interfaces in Crystalline Materials,* Oxford University Press, New York.

Wolf, D. and Yip, S., Eds. (1992) *Materials Interfaces, Atomic-level Structure and Properties,* Chapman and Hall, New York.

Specific References

Anstis, G.R. and Cockayne, D.J.H. (1979) *Acta Cryst.* **A35**, 511.
Doyle, P.A. and Turner, P.S. (1968) *Acta Cryst.* **A24**, 390.
Gevers, R., Art, A., and Amelinckx, S. (1963) *Phys. Stat. Sol.* **3**, 1563.
Gevers, R., Blank, H., and Amelinckx, S. (1966) *Phys. Stat. Sol.* **13**, 449.
Hashimoto, H., Howie, A., and Whelan, M.J. (1962) *Proc. Roy. Soc. London* **A269**, 80.
Hirsch, P., Howie, A., Nicholson, R.B., Pashley, D.W., and Whelan, M.J. (1977) *Electron Microscopy of Thin Crystals,* 2nd edition, p. 225, Krieger, Huntington, New York.
Howie, A. and Basinski, Z.S. (1968) *Phil. Mag.* **17**, 1039.
Howie, A. and Sworn, H. (1970) *Phil. Mag.* **31**, 861.
Humphreys, C.J. and Hirsch, P.B. (1968) *Phil. Mag.* **18**, 115.
Mott, N.F. and Massey, H.S.W. (1965) *The Theory of Atomic Collisions,* 3rd edition, Clarendon Press, Oxford.
Rasmussen, D.R., McKernan, S., and Carter, C.B. (1991) *Phys. Rev. Lett.* **66**, 2629.
Schäublin, R. and Stadelmann, P. (1993) *Mater. Sci. Engng.* **A164**, 373.
Taftø, J. and Spence, J.C.H. (1982) *J. Cryst.* **15**, 60.
Thölen, A.R. (1970) *Phil. Mag.* **22**, 175.
Thölen, A.R. (1970) *Phys. Stat. Sol. (A)* **2**, 537.
Viguier, B., Hemker, K.J., and Vanderschaeve, G. (1994) *Phil. Mag.* **A69**, 19.
Yoshioka, H. (1957) *J. Phys. Soc. Japan* **12**, 618.

Strain Fields

25

CHAPTER PREVIEW

25.1. Why Image Strain Fields? 403
25.2. Howie–Whelan Equations 403
25.3. Contrast from a Single Dislocation 405
25.4. Displacement Fields and Ewald's Sphere 408
25.5. Dislocation Nodes and Networks 409
25.6. Dislocation Loops and Dipoles 409
25.7. Dislocation Pairs, Arrays, and Tangles 411
25.8. Surface Effects 412
25.9. Dislocations and Interfaces 413
25.10. Volume Defects and Particles 417
25.11. Simulating Images 418
    25.11.A. The Defect Geometry 418
    25.11.B. Crystal Defects and Calculating the Displacement Field 418
    25.11.C. The Parameters 419

CHAPTER PREVIEW

As we discussed in Chapter 23, bending of the lattice planes causes a change in the diffraction conditions and therefore a change in the contrast of the image. The presence of a lattice defect in the specimen causes the planes to bend close to the defect. The special feature here is that the bending varies not just laterally, but also through the specimen. Since the details of the bending generally depend on the characteristics of the defect, we can learn about the defect by studying the contrast in the TEM image. This simple principle has led to one of the main applications of TEM, namely, the study of defects in crystalline materials. We can claim that our understanding of the whole field of dislocations and interfaces, for example, has advanced because of TEM. We have even discovered new defects using TEM.

    Usually we want to learn two things about these defects: we want to know where they are and then understand what they are. So the idea underlying this chapter is the same as for bend contours: we use different reflections corresponding to different sets of lattice planes. We see how the defects affect the image contrast from those different lattice planes and thus characterize the defects. In case you are worried, we would like to emphasize that this is *not* a chapter about defects, it is concerned with understanding contrast in the TEM. We will introduce the necessary terminology and notation concerning defects, but we won't try to give you a comprehensive discussion of them. You should consult the standard references on dislocations at the end of the chapter if you need more details. However, we will show lots of pictures because now we are concerned with the appearance of images.

Strain Fields

25.1. WHY IMAGE STRAIN FIELDS?

First, we should review our terminology. When we displace the atom at position **r** a distance **R(r)** from its site in the perfect crystal, we say the crystal is under a strain (ε). If the crystal is strained then it must be subject to a stress, which we'll call σ. (Metallurgists traditionally use these symbols and although σ means "cross section" to a microscopist, we'll stick with it.) Since **R(r)** varies with position in the crystal, ε and σ will in general also vary with **r**. We will assume that ε and σ can each be defined at a point. Then we will refer to these quantities as the displacement field, **R(r)**, the strain field, ε(**r**), and the stress field, σ(**r**). You will notice that these terms are used interchangeably in the literature. What we image is the effect of the **R(r)**.

To have an intuitive feel for why we see contrast from dislocations, consider the geometry shown in Figure 25.1. The diffraction geometry has been set up so that the specimen is slightly tilted away from the Bragg condition. The distortion due to the dislocation will then bend the near-diffracting planes back into the Bragg-diffracting condition. We have relrods so there will still be some intensity in the electron beam even when we are not at the exact Bragg condition. The figure shows planes bending at a dislocation; compare this to Figure 23.7 showing bend contours. Regions far from the dislocation are tilted well away from the Bragg condition, while the regions either side of the dislocation core are at the Bragg condition for $\pm \mathbf{g}_{hk\ell}$. It is more difficult to "see" the diffracting planes for a screw dislocation (Figure 25.2A) but the effect is the same.

When studying a particular dislocation (edge or screw), we want to determine the following parameters:

- The direction and magnitude of the Burgers vector, **b**, which is normal to the $hk\ell$ diffracting planes (Figures 25.1 and 25.2B).
- The line direction (a vector) and therefore the character of the dislocations (edge, screw, or mixed).
- The glide plane.

There are other questions we want to answer:

- Is the dislocation interacting with other dislocations, or with other lattice defects?
- Is the dislocation jogged, kinked, or straight?
- What is the density of dislocations in that region of the specimen (and what was it before we prepared the specimen)?
- Has the dislocation adopted some special configuration, such as a helix?

In many of these questions, you may find that stereomicroscopy (Section 31.1) can be very helpful, although we will not emphasize that technique. The basic requirement if you do use stereomicroscopy is that you must form all of your images using the same **g**-vector.

25.2. HOWIE–WHELAN EQUATIONS

Let's start with the two-beam phenomenological approach because it worked so well in Chapter 24. An important assumption is that we have linear elasticity. What this means is that if we have $\mathbf{R}_1$ due to one defect and $\mathbf{R}_2$ due to a second defect, then at any point in the specimen we can just add these two values to determine the total displacement field, **R**. We will not consider anisotropic elasticity, although this can readily be included in calculations.

In Chapter 24, we showed that we could modify the Howie–Whelan equations to include a lattice distortion **R**. So for the imperfect crystal

$$\frac{d\phi_g}{dz} = \frac{\pi i}{\xi_0}\phi_g + \frac{\pi i}{\xi_g}\phi_0 \exp\left[-2\pi i(sz + \mathbf{g}\cdot\mathbf{R})\right] \quad [25.1]$$

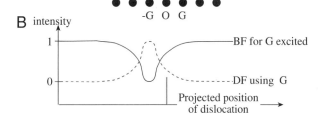

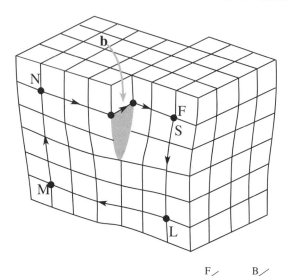

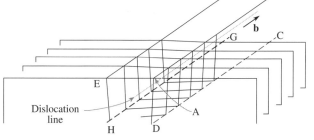

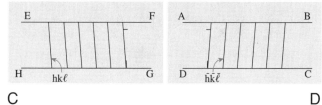

Figure 25.1. (A) The specimen is tilted slightly away from the Bragg condition ($s \neq 0$). The distorted planes close to the edge dislocation are bent back into the Bragg-diffracting condition ($s = 0$), diffracting into G and –G as shown. (B) Schematic profiles across the dislocation image showing that the defect contrast is displaced from the projected position of the defect.

Now we make a different substitution of variables (compare with equations 24.5 and 24.6). Set

$$\phi_0(z)_{(sub)} = \phi_0(z) \exp\left(\frac{-\pi i z}{\xi_0}\right) \qquad [25.2]$$

and

$$\phi_g(z)_{(sub)} = \phi_g \exp\left(2\pi i s z - \frac{\pi i z}{\xi_0} + 2\pi i \, \mathbf{g} \cdot \mathbf{R}\right) \qquad [25.3]$$

The justification for this substitution is the same as always. You'll notice that $\phi_0(z)_{(sub)}$ is the same as before, but $\phi_{g(sub)}$ now includes a $\mathbf{g} \cdot \mathbf{R}$ term. The reason for this substitution is that it will give us a simple expression for $d\phi_g/dz$.

The equations become

$$\frac{d\phi_0(z)_{(sub)}}{dz} = \frac{\pi i}{\xi_g} \phi_g(z)_{(sub)} \qquad [25.4]$$

and

$$\frac{d\phi_{g(sub)}}{dz} = \frac{\pi i}{\xi_g} \phi_0(z)_{(sub)} + \left[2\pi i \left(s + \mathbf{g} \cdot \frac{d\mathbf{R}}{dz}\right)\right] \phi_g(z)_{(sub)} \qquad [25.5]$$

Figure 25.2. (A) Distortion of planes around a screw dislocation. The circuit SLMNF is used to define the Burgers vector, **b** (see Figure 25.5). (B) Schematic showing the rotation of the diffracting planes by a screw dislocation. The planes are rotated in opposite directions on either side (C, D) of the dislocation.

which can be rewritten, while dropping the subscript

$$\frac{d\phi_g}{dz} = \frac{\pi i}{\xi_g} \phi_0 + 2\pi i s_R \phi_g \qquad [25.6]$$

This equation looks just like equation 13.14 but with s_R instead of s, where

$$s_R = s + \mathbf{g} \cdot \frac{d\mathbf{R}}{dz} \qquad [25.7]$$

The concept of s_R is new.

The importance of this result is that although we have a new "s," we have the same equation so we can use the rest of the analysis of Chapter 13 and obtain the same results with a modified value of s, i.e., s_R. Therefore, we'll have the same thickness dependence so that the contrast of the defects will depend on both s and ξ_g. The big change is that we can now treat the case where $\mathbf{R}$ is a continuous function of z.

We will examine how the $\mathbf{g} \cdot d\mathbf{R}/dz$ and $\mathbf{g} \cdot \mathbf{R}$ terms are used to understand dislocations. Since the equations we have just derived have the same form as those we discussed in Chapters 11 and 24, we can expect many of the same properties in the images. In particular, the images of defects will show the same sort of thickness dependence. We can also use the equations we derived in Chapter 24, so we have two ways of looking at the defects:

- $\mathbf{g} \cdot \mathbf{R}$ contrast is used when $\mathbf{R}$ has a single value,
- s_R contrast is used when $\mathbf{R}$ is a continuously varying function of z,

which in turn is associated with $\mathbf{g} \cdot d\mathbf{R}/dz$.

Now let's consider the principles of this analysis. Remember, we are not trying to be quantitative or totally rigorous. We will generalize the two-beam treatment for the imperfect crystal. Note that we still have beams, it's a dynamical situation, and we assume that the column approximation is valid (Hirsch *et al.* 1960). So how does the column approximation relate to the theory? The model relates $\mathbf{R}$ to the column, as shown in Figure 25.3, and the calculation is for a continuum even though we have atoms. The important point is that for a displacement field, $\mathbf{R}$ varies with position, $\mathbf{r}$; we can define the origin as the core of the defect. We'll go through the calculation for a dislocation parallel to the foil surface.

As we saw in Section 13.11, the column approximation is equivalent to assuming that the crystal can be divided into narrow columns. We then calculate the amplitudes of the beams in any such column as if the whole crystal consisted of an infinite number of identical columns. The approximation is valid when we don't need to see image detail below ~2–3 nm. The actual diameter of the column depends on the diffracting conditions (Howie and Basinski 1968; Howie and Sworn 1970). We can include the effect of distortions due to strains from lattice defects by imagining that the column consists of slabs of perfect crystal each displaced by an amount $\mathbf{R}(z)$ (see Section 24.13). Remember that z is actually measured along the column.

25.3. CONTRAST FROM A SINGLE DISLOCATION

When we study dislocations, we usually want to know how many there are (the density) and whether they are edge, screw, or mixed in character. The displacement field in an isotropic solid for the general, or mixed, case (Hirth and Lothe 1982) can be written as

$$\mathbf{R} = \frac{1}{2\pi}\left(\mathbf{b}\phi + \frac{1}{4(1-v)}\{\mathbf{b}_e + \mathbf{b} \times \mathbf{u}(2(1-2v)\ln r + \cos 2\phi)\}\right) \quad [25.8]$$

For convenience, $\mathbf{R}$ is given here in polar coordinates (r and ϕ) shown in Figure 25.3; $\mathbf{b}$ is the Burgers vector, $\mathbf{b}_e$ is the edge component of the Burgers vector, $\mathbf{u}$ is a unit vector along the dislocation line (the line direction), and v is Poisson's ratio.

It was particularly important to be able to write down this expression when we did the calculations by hand. However, when we have a computer available, it's quite straightforward to use anisotropic elasticity (Steeds 1973) or just feed in displacements calculated from a computer model of the atom structure.

The amplitude of the diffracted beam, ϕ_g, is directly influenced by the value of $\mathbf{R}$. We can consider two particular cases, namely, the screw and edge dislocations. For the screw dislocation, $\mathbf{b}_e = 0$ and $\mathbf{b}$ is parallel to $\mathbf{u}$ so that $\mathbf{b} \times \mathbf{u} = 0$. Then the expression for $\mathbf{R}$ in equation 25.8 simplifies to

$$\mathbf{R} = \mathbf{b}\frac{\phi}{2\pi} = \frac{\mathbf{b}}{2\pi}\tan^{-1}\left(\frac{z-z_d}{x}\right) \quad [25.9]$$

Here, z is the distance traveled down the column and z_d is the distance of the dislocation core below the top surface (again, refer to Figure 25.3). The dependence on $(z - z_d)$ emphasizes that the displacement field is present above

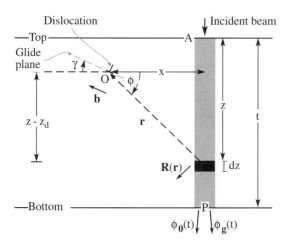

Figure 25.3. The effect of a dislocation with Burgers vector, $\mathbf{b}$, at O on a column, distance x away. The effect of the strain field on the electron waves in the column is integrated in increments dz over its total length t, giving amplitudes $\phi_0(t)$ and $\phi_g(t)$ at P.

Table 25.1. Different Burgers Vectors and Different Reflections Give Different g·b = n Values[a]

g \ b	$\frac{1}{6}[11\bar{2}]$	$\frac{1}{6}[1\bar{2}1]$	$\frac{1}{6}[\bar{2}11]$	$\frac{1}{3}[111]$
$\pm(1\bar{1}1)$	$\pm 1/3$	$\pm 2/3$	$\pm 1/3$	$\pm 1/3$
$\pm(\bar{1}\bar{1}1)$	$\pm 2/3$	$\pm 1/3$	$\pm 1/3$	$\pm 1/3$
$\pm(0\bar{2}2)$	± 1	± 1	0	0
$\pm(200)$	$\pm 1/3$	$\pm 1/3$	$\pm 2/3$	$\pm 2/3$
$\pm(3\bar{1}1)$	0	± 1	± 1	± 1
$\pm(\bar{3}\bar{1}1)$	± 1	0	± 1	± 1

[a] The dislocations all lie on a (111) plane in an fcc material; the beam direction is [011].

and below the dislocation; it affects the whole column. From these two equations we see that **g·R** is proportional to **g·b**. For this reason, we often discuss images of dislocations in terms of **g·b** (g-dot-b) contrast. Examples of **g·b** values for some dislocations lying on a (111) plane in an fcc material with a [011] beam direction are given in Table 25.1.

The second special case arises when the dislocation is pure edge in character. Then $\mathbf{b} = \mathbf{b}_e$ and **g·R** involves two terms, **g·b** and **g·b×u**. (The latter term is read as "g-dot-b-cross-u.") The displacement field causes the Bragg-diffracting planes associated with **g** to bend. Incidentally, the origin of **g·b×u** is interesting; it arises because the glide plane is buckled by the presence of an edge dislocation (Hirth and Lothe 1982) as illustrated in Figure 25.4. This buckling can be important because it complicates the analysis of **b** for some dislocations with an edge component, as we'll see below.

- Always remember: **g·R** causes the contrast and for a dislocation, **R** changes with z.
- We say that **g·b** = n. If we know **g** and we determine n, then we know **b**.

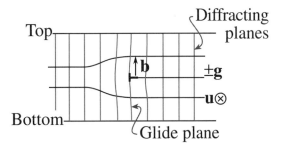

Figure 25.4. Buckling of the glide planes arises because of the term **g·b×u** and is important because it complicates the analysis of **b**.

An experimental point: you usually set s to be greater than 0 when imaging a dislocation. Then the dislocation can appear dark against a bright background in a BF image. Of course, you still need to think about s_R and $d\mathbf{R}/dz$ since these will vary with z, as we saw in Figure 25.1.

The + and – signs in Table 25.1 are very important. If the sign of **R**, and hence **g·R** or **g·b**, reverses, then the image of the dislocation will move to the other side of the projected position of the dislocation core. If you look carefully at Figure 25.1, you can appreciate that reversing the sign of s produces the same effect as reversing the sign of **g**. We can summarize these two ideas in terms of the quantity (**g·b**)s ("g-dot-b-times-s"), as shown in Figure 25.5.

> If **g·b** = 0, then you won't see any contrast because the diffracting planes are then parallel to **R**. This is termed the *invisibility criterion*.

If we identify two reflections, $\mathbf{g}_1$ and $\mathbf{g}_2$, for which **g·b** = 0, then $\mathbf{g}_1 \times \mathbf{g}_2$ is parallel to **b**. This identification of **b** is actually a little more complicated because dislocations appear out of contrast when **g·b** < 1/3; similarly, dislocations need not be invisible even if **g·b** = 0 when **g·b×u** ≠ 0. Further exceptions to the rule are given in Edington (1976); see also Section 25.15.

If we compare the contrast from a dislocation with that from a SF, the difference is that now α is a continuously varying function of z. The image of the dislocation itself shows thickness fringes, but it may be "out of contrast" at some depths or thicknesses, as you can see in the experimental image shown in Figure 25.6A.

Some points to remember from this discussion are:

- The sign of s affects the image.
- The sign of x affects the image; the image is asymmetric.
- The magnitude of **s** affects the image.
- The depth of the dislocation and the thickness of the specimen affect the image.
- The appearance of the image depends on **g·b** or, more completely, on (**g·b**)s and **g·b×u**.
- If we repeat this analysis for other values of **g·b** (= n) and plot intensities, we would find that the image width becomes broader as n increases.
- Note where the dislocation image "comes from": the position of the line in the image only rarely corresponds to the projected position of the dislocation; it is usually displaced to one side of the core.
- As a complication, remember that the dislocations will probably be found in wedge specimens, not ideal parallel-sided ones.

25 ■ STRAIN FIELDS

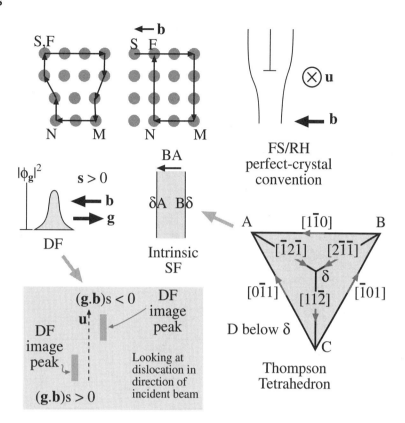

Figure 25.5. A brief summary of dislocations in an fcc crystal: **b** is defined by the finish- (F) to-start (S) vector in a right-hand (RH) circuit that comes to closure around the dislocation but fails to close in the perfect crystal. The location of the diffracted intensity $|\phi_g|^2$ relative to the core depends on the sign of **b**, **g**, and s for the FSRH convention. If any sign is reversed, the contrast shifts across the core. When a perfect dislocation splits into Shockley partial dislocations, the order of the partial dislocations is given by the Thompson tetrahedron.

A final "rule of thumb" which you may find useful (from computer modeling and early analytical calculations) is

> If **g·b** = 0, you can still "see" dislocations when **g·b×u** ≥ 0.64. For fcc materials, this rule can be useful when the foil is not parallel to a {111} plane.

Other examples of dislocation images are illustrated in Figure 25.6. Remember that partial dislocations are not only present in fcc metals; they also occur in many fcc semiconductors and many layer materials. Such materials may have a very low stacking-fault energy allowing the partial dislocations to separate, forming wide ribbon-like defects, as shown in Figures 25.6A–C. The single line below the arrow in (C) is a dislocation having its Burgers vector parallel to **g** here (so **g·b** = 2); you can see two "peaks" in the image, one darker and broader than the other. Notice that one peak has nearly disappeared in (B) and the broad peak is on the other side of the dislocation. A group of three parallel lines is present in (C) but is out of contrast in (B). These are three Shockley partial dislocations all having the same **b** and thus all giving **g·b** = 0 in (B) (the three lines actually form by the dissociation of a perfect dislocation with Burgers vector $\frac{1}{2}$<112>). The 11$\bar{1}$ image (A) is formed by tilting the specimen to a 112 pole (~20° from the 111 pole) and shows contrast from the stacking faults themselves; these faults will never give contrast at the 111 pole since **g·R** is then always 0 (or an integer). The intermetallics tend to have large unit cells so that the superlattice dislocations dissociate into partial dislocations, which would have been perfect dislocations in the disordered crystal (Figures 25.6D and E). These super-partial dislocations can dissociate further as they might have in the disordered lattice, or they can separate differently in different ordered domains (Figure 25.6F). Dislocations in interphase boundaries can be revealed by imaging with different reflections (Figures 25.6G and H). Since the dislocations are present to accommodate the mismatch, they must lie at, or close to, the (001) phase boundary. It can be difficult to analyze their Burgers vectors unambiguously, because the adjoining materials have different extinction distances, etc. One of the extra challenges is determining the plane on which this dissociation occurs. We'll illustrate how we can see **g·b×u** contrast when we examine dislocation loops in Zn in Section 25.6.

25.4. DISPLACEMENT FIELDS AND EWALD'S SPHERE

In Section 25.3, we showed that when a displacement $\mathbf{R}(\mathbf{r})$ is present, we can think of s as being replaced by $s_\mathbf{R}$ (equation 25.7). Hirsch et al. (1977) (see also Goringe 1975 and Section 25.13) showed that this new s should be written more completely as

$$s_\mathbf{R} = s + \mathbf{g} \cdot \frac{\partial \mathbf{R}}{\partial z} + \theta_B \mathbf{g} \cdot \frac{\partial \mathbf{R}}{\partial x} \qquad [25.10]$$

The point is that, as you can see in Figure 25.7, $\mathbf{R}$ causes the lattice planes to bend through an angle $\delta\phi$. So two other parameters, namely, $\mathbf{g}$ and s, also change. The diffraction vector is actually lengthened by $\Delta\mathbf{g}$ but, more importantly, $\mathbf{g}$ is rotated. The result is that s increases by the two com-

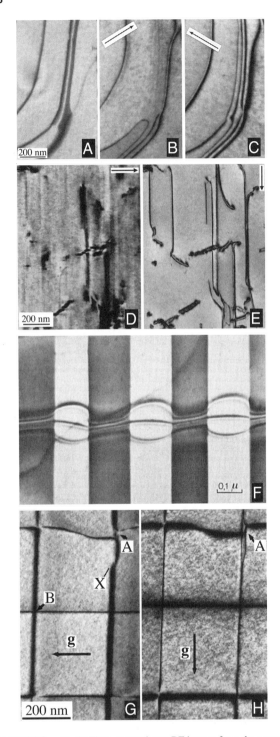

Figure 25.6. (A–C) Three strong-beam BF images from the same area using (A) $\{11\bar{1}\}$ and (B,C) $\{220\}$ reflections to image dislocations which lie nearly parallel to the (111) foil surface in a Cu alloy which has a low stacking-fault energy. (D,E) Dislocations in Ni$_3$Al in a (001) foil imaged in two orthogonal $\{220\}$ reflections. Most of the dislocations are out of contrast in (D). (F) A complex dislocation crossing a (rotational) domain boundary; the character of the dislocation changes and thus its dissociation width changes. (G,H) Dislocations in a (001) interface between two slightly lattice-mismatched III-V compounds.

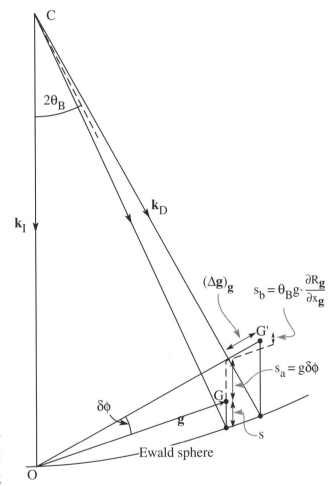

Figure 25.7. The strain field of the dislocation causes the lattice planes to bend through an angle $\delta\phi$. So $\mathbf{g}$ and s also change. The diffraction vector is lengthened by $\Delta\mathbf{g}$ and $\mathbf{g}$ is rotated. So s increases by the two components of $s_\mathbf{R}$, i.e., s_a and s_b.

ponents, s_a and s_b, shown in the figure, to give s_R. If you manipulate this equation for small angles you produce equation 25.8. We usually neglect the third term because θ_B is small, but it can become important when screw dislocations intersect the surface (Tunstall *et al.* 1964).

An alternative way of looking at this deformation is to think of **g** as changing by Δ**g**. We can define this change by the equation

$$\mathbf{g} \cdot (\mathbf{r} - \mathbf{R}(\mathbf{r})) = (\mathbf{g} + \Delta\mathbf{g}) \cdot \mathbf{r} \quad [25.11]$$

so that

$$-\mathbf{g} \cdot \mathbf{R}(\mathbf{r}) = \Delta\mathbf{g} \cdot \mathbf{r} \quad [25.12]$$

The implication is that the information about the displacement field, $\mathbf{R}(\mathbf{r})$, is present in the region around **g** but not actually at **g**. Remember that the reflection **g** is present because we have a perfect crystal. It is difficult to image this type of scattering. If you displace the objective aperture, you will still see the dislocation, but other inelastic scattering will complicate image interpretation. We saw that scattering does indeed occur between Bragg reflections in Section 17.6. An analogy for scattering from dislocations is the scattering of light from a single slit which we discussed in Chapter 2.

In the deformable-ion approximation (Section 24.13), we make the assumption that the atom doesn't know it has moved. If $\mathbf{R}(\mathbf{r})$ varies rapidly, as it does near the core of a dislocation, the approximation must fail. You can draw the same conclusion whenever the density of the material changes rapidly. So what we should do is use a better model for the atomic potential, one that also takes account of what happens to the valence electrons at such a defect. Of course, linear elasticity theory also fails when the strains, and hence $\mathbf{R}(\mathbf{r})$, are large, as at dislocation cores.

25.5. DISLOCATION NODES AND NETWORKS

You can analyze the Burgers vectors of dislocations which form networks directly and easily, if all the dislocations lie in a plane parallel to the surface of the specimen, as illustrated in Figure 25.8 for the case of graphite. The idea is simple: you form a series of images using different **g**-vectors. Don't forget that you can tilt to other poles; in fact, you'll often need to tilt the specimen just to image SFs which lie parallel to the foil surface, as in Figure 25.8A. Such tilting experiments are essential if you're examining networks of misfit dislocations, since the dislocations will

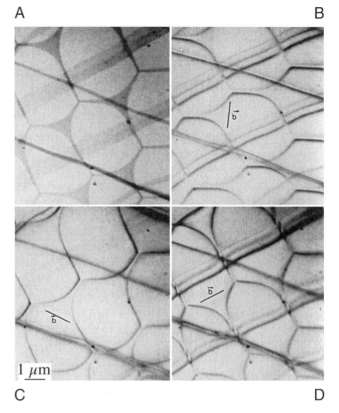

Figure 25.8. Dislocation networks in graphite. In (A) the stacking faults parallel to the foil surface are imaged, in (B), (C), and (D) two of the three dislocations at each node are in contrast but the third is invisible. Knowing **g** for each image, the Burgers vector of the dislocations can be determined as shown.

then often have a component of their Burgers vector out of the plane of the network.

25.6. DISLOCATION LOOPS AND DIPOLES

Loops have been studied extensively because they can form when point defects coalesce. There are probably thousands of papers describing TEM studies of radiation damage and the formation of dislocation loops. In fact, many HVEMs were built in the 1960s just to study this problem. Questions which were answered led to a greatly improved understanding of irradiation processes (but failed to justify the construction of more nuclear power stations). We found that:

- The loops can form by coalescence of interstitials or vacancies.
- The rate of growth, critical size, and nucleation time for different loops can be measured.

- Some of the loops are faulted (containing a SF) while others are not faulted. The faulting should be related to the size of the loop and the stacking-fault energy of the material.

These studies were particularly instructive illustrations of the value of diffraction contrast.

- Dislocation loops can have either positive or negative **b**, as shown in Figure 25.9.
- Loops can be present which show no **g·b** contrast.
- Loops can enclose single or multiple stacking faults, and so exhibit SF contrast as shown in Figure 25.10.
- The dislocation dipole is a special case and gives an important example of interacting dislocations. TEM is the best way to image

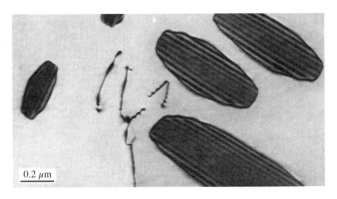

Figure 25.10. Dislocation loops in irradiated Ni showing SF contrast.

dipoles because they have no long-range strain fields; the Burgers vector of the complete dipole is zero!

Dislocations in Zn provide a particularly nice illustration of **g·b×u** contrast. If the specimen surface is parallel to the (0001) basal plane, then dislocation loops can readily form by coalescence of vacancies. In Figure 25.11, **b** is normal to **g** so that **g·b** = 0. These loops give a clear illustration of how the appearance of the image depends on the line direction, **u**, of a dislocation. Note that you can see the dislocation, even though **g·b** is zero, so this is not an absolute criterion for invisibility.

The above discussion is fine if the loops are large, but a problem arises when they are small. You must then consider the details of the contrast mechanism.

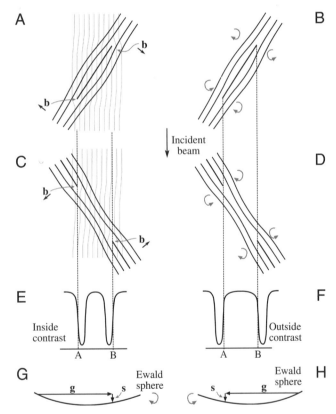

Figure 25.9. (A) Structure of an interstitial loop relative to the diffracting planes (faint lines). (B) Arrows show the rotation of the diffracting planes around the dislocation. (C,D) Vacancy loops. (E,F) Position of the image contrast relative to the projected dislocation position. Inside contrast occurs when clockwise rotation of the diffracting planes brings them into the Bragg condition. Outside contrast occurs for the counter-clockwise case. (G,H) The relationship between **g**, **s** and the sense of rotation. Everything is reversed if the loops are tilted in the opposite direction relative to the beam (i.e., reflect this figure in a mirror).

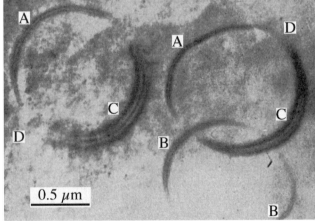

Figure 25.11. Prismatic loops in Zn parallel to the (0001) surface of the specimen with **b** = c[0001]. All round the loop, **b** is normal to **g** so that **g·b** = 0 and the vector **b×u** lies in the plane of the loop. At A, B, and C, **b×u** is parallel to **g** so that we see strong contrast. However, at D, **b×u** and **g** are mutually perpendicular so that **g·b×u** = 0 and the loop disappears.

> The basic idea is that the appearance of the image is now dependent on the thickness of the specimen.

This was, of course, true for all these images, but now the size of the defect is small compared to the extinction distance. The schematic shown in Figure 25.12 summarizes the contrast which arises from small vacancy loops; if the loops were interstitial in nature, the contrast would be reversed. Not only does the black/white contrast change as the position of the defect changes in the specimen, but its size also appears to change. When the nature of the loops becomes more complex, the appearance of the image may also become more difficult to interpret with "butterflies," "lozenges," and "peanuts" being common terms. Notice that the behavior of the contrast differs in BF and DF images; this effect is similar to that which we discussed in Chapter 24 and is again related to anomalous absorption. A detailed description of this complex contrast behavior is given by Wilkens (1978).

Dislocation dipoles can be present in great numbers in heavily deformed metals, but can also be important in the degradation of some semiconductor devices. Dipoles can be thought of as loops which are so elongated that they look like a pair of single dislocations of opposite Burgers vectors, lying on parallel glide planes. As a result, they are

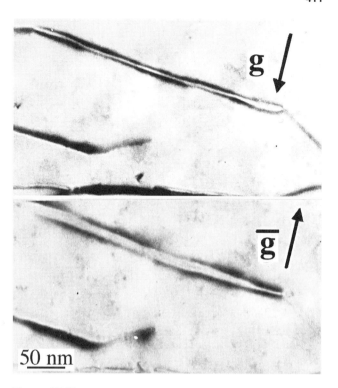

Figure 25.13. Images of dislocation dipoles in Cu showing inside–outside contrast on reversing **g** (±220).

best recognized by their "inside–outside" contrast as illustrated in Figure 25.13. You can appreciate the origin of the term by looking at the projection of the images of the two dislocations when you reverse the sign of **g**: since the two dislocations have opposite Burgers vectors, Figure 25.9 tells you that one image will lie on one side of the core and the other on the opposite side. The order reverses when you reverse **g**.

25.7. DISLOCATION PAIRS, ARRAYS, AND TANGLES

Remember, you are not limited to **g**-vectors which are parallel to the foil surface; hence you can tilt the specimen to see SF contrast. As we saw in Figure 25.8, this is often helpful if you have SFs associated with the dislocations; you can then produce **g·R** contrast for the fault. We will discuss dislocation dissociation more in Chapter 26. If you look back at Figure 25.6, you will see the benefit of being able to see the SF. This figure also illustrates the effect of n on the dislocation contrast.

Consider a dislocation in an fcc metal which can dissociate into two Shockley partial dislocations on the (111) plane. We can write down the dislocation reaction as

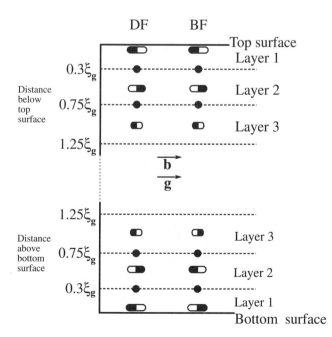

Figure 25.12. Changes in the black–white contrast from small dislocation loops at different layer depths in the specimen. The DF shows the same contrast at the top and bottom while the BF contrast is complementary at the two surfaces.

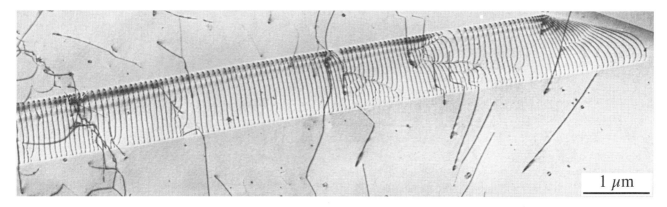

Figure 25.14. Dislocations threading through a very thick specimen in an image recorded using a high-voltage TEM.

$$\frac{1}{2}\left[1\bar{1}0\right] = \frac{1}{6}\left[1\bar{2}1\right] + \frac{1}{6}\left[2\bar{1}\bar{1}\right] \quad \text{on} \left(111\right) \quad [25.13]$$

If we image this dislocation using the $(2\bar{2}0)$ reflection, then $\mathbf{g}\cdot\mathbf{b} = 2$. If, instead, we use the $(20\bar{2})$ reflection, then $\mathbf{g}\cdot\mathbf{b} = 1$. The appearance of the image is very different even if we cannot see the individual partial dislocations.

The advantage of using high voltages to study arrays of dislocations is illustrated in Figure 25.14; everything we said in Chapter 11 applies when we study dislocations. We see thickness fringes at the surface, but these disappear in the central region of the foil. When the foils are this thick, you may find stereomicroscopy helpful in giving a 3D view of the defect arrangement; you can imagine its value in interpreting an image such as that shown in Figure 25.15A. The defects may be very close together in heavily deformed materials, as shown in Figure 25.15B; you then need to make the specimen thin over very large areas to minimize image overlap (e.g., Hughes and Hansen 1995). If the density of defects is too large, the weak-beam technique may be the only way to "look into" the walls (Chapter 26).

25.8. SURFACE EFFECTS

In TEM, we always have thin foils. Dislocation strain fields are long range, but we often assign them a cut-off radius of ~50 nm. However, the thickness of the specimen might only be 50 nm or less, so we can expect the surface to affect the strain field of the dislocation, and vice versa.

When an edge dislocation lies parallel to the surface of a very thin specimen, it causes the specimen to bend. The effect is not large, but large enough compared to the Bragg angle, as illustrated schematically and with an example in Figure 25.16 (Amelinckx 1964).

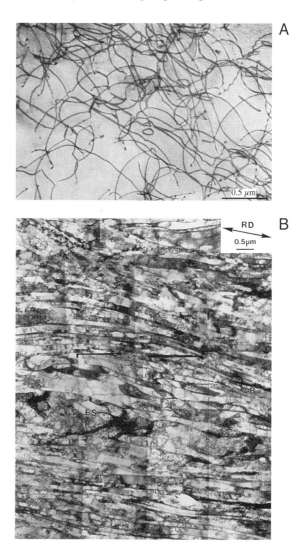

Figure 25.15. (A) Dislocation tangles in an Fe-35% Ni-20% Cr alloy, creep tested at 700°C; the dislocations have moved by glide and climb and do not lie on well-defined planes. (B) Dislocation walls in Al which has been heavily deformed by directional rolling.

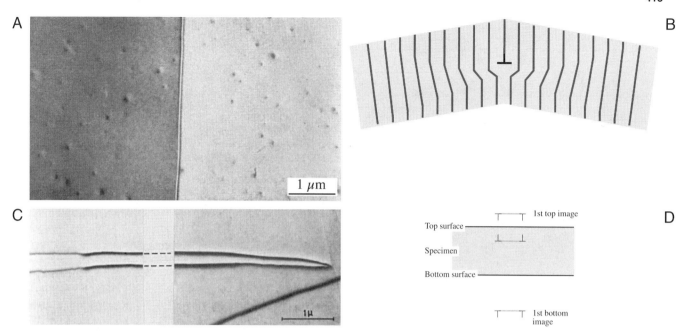

Figure 25.16. (A) A single-edge dislocation lying parallel to the surface of a very thin foil of SnSe$_2$ causes the diffracting planes to bend (B) so we see different intensity in the matrix on either side of the defect. (C,D) If the dislocation is dissociated, the image forces due to the surface cause its width to decrease. The schematic in (D) shows the image dislocations included to represent the effect of the surface.

A calculation: $\sin \theta_D/2 = \mathbf{b}/2t$. If $\mathbf{b} \sim 0.25$ nm and $t = 50$ nm, $\sin \theta_D/2 = 0.0025$ and $\theta_D = 0.29°$, which you can compare to a Bragg angle from $\sin \theta_B = (n\lambda/2d)$ $(0.0037 \text{ nm}/0.5 \text{ nm} = 0.0074)$. $\theta_B = 0.42°$. Notice how θ_D increases for thinner foils and θ_B decreases for increasing voltage (decreasing λ).

Similarly, if the dislocation is dissociated, the proximity of the surface causes its width to decrease. We can model this situation using "image dislocations" as shown in Figures 25.16C and D. The main point is that we can think of these image dislocations as forcing the partial dislocations closer together; the proximity of the surface can really change the structure of the defect, not just its contrast. A similar effect can occur when the dislocation is inclined to the surface and can result in a V-shaped geometry, as we'll show in Section 26.8.

A special interaction between dislocations and surfaces occurs when a dislocation tries to glide out of the material but can't penetrate a surface layer (which might even be amorphous as in the case of oxide films on metals), as shown in Figure 25.17.

Takayanagi (1988) has shown that we can have dislocations at the surface just because the structure of the surface layer is different from that of the bulk material. The surface of materials can actually reconstruct. The surface of a (111) Au film is more dense than the rest of the film and this misfit is accommodated by surface dislocations, as shown in Figure 25.18. We see the contrast because the strain field extends into the bulk layer. The identification of these dislocations has been confirmed using STM, which also gives more information on the detailed surface structure. However, they were observed first by TEM. The difficulty in TEM studies is that the surface contaminates unless you operate under UHV conditions.

Dislocation can be viewed nearly parallel to their line directions, when we still see contrast even for screw dislocations, as you can see in Figure 25.19A (Tunstall *et al.* 1964). Initially, this contrast is surprising since **g·b** and **g·b×u** must be zero for any screw dislocation. However, the screw dislocation can relax at the surface, as shown in Figure 25.19B (Amelinckx 1992).

25.9. DISLOCATIONS AND INTERFACES

Interfaces are, of course, important in all polycrystalline materials. In metals, semiconductors, and thin films on substrates the interaction between dislocations and interfaces is critical. So now we'll briefly examine the special features we see when combining line and planar defects as illustrated in Figure 25.20. This is one topic where image simulation, which we'll discuss in Section 25.12, is invaluable.

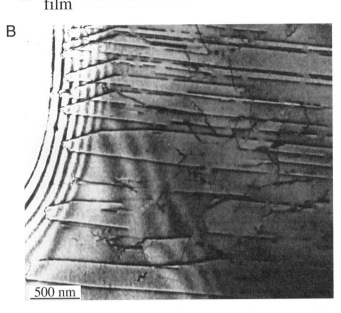

Figure 25.17. (A) Schematic diagram of dislocations pinned at the surface of the specimen by surface films such as oxides. (B) A reduced (i.e., metal) film on NiO pins dislocations. Such films may be introduced during or after thinning to electron transparency.

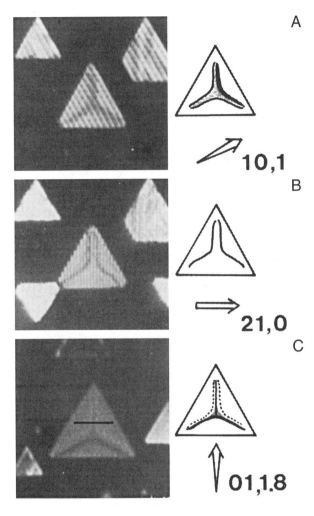

Figure 25.18. Dislocations networks can form at the surface of (111) Au islands because the surface layer relaxes to a "lattice" parameter which is different from that of the bulk material. Different dislocations are visible under different diffracting conditions (A–C).

When we have an array of dislocations, the strain fields overlap so that the value of **R(r)** for each dislocation tends to be reduced. This is the grain boundary model of an interface.

Dislocations can be present at interfaces where the composition, or structure, or both change.

- Misfit dislocations accommodate the difference in lattice parameter between two well-aligned crystalline grains. Surface dislocations (as we saw in Section 25.8) are a special subgroup of misfit dislocations.
- Transformation dislocations are the dislocations which move to create a change in orientation or phase. The $\frac{1}{6}$<112> dislocations in twin boundaries in fcc materials are an example of transformation dislocations (twinning dislocations).

A complication in the analysis of images of interfacial dislocations is that they are often associated with steps in the interface. An example of such steps is shown in Figure 25.21. Sometimes, as is the case for the $\frac{1}{6}$<112> twinning dislocations, the dislocations must introduce a step. In other situations, steps are present but there is no dislocation. The difficulty is that we often encounter all three of these situations at the same time. We will also examine these defects, using weak-beam conditions in Chapter 26 and using HRTEM in Chapter 28.

We will discuss the images first and then, remembering that information must also be present in the DP, we will relate the two.

In many cases that interest us, grain boundaries appear as arrays of dislocations. In general, the grains are misoriented. There are some special cases, as we saw in Chapter 24.

25 ■ STRAIN FIELDS

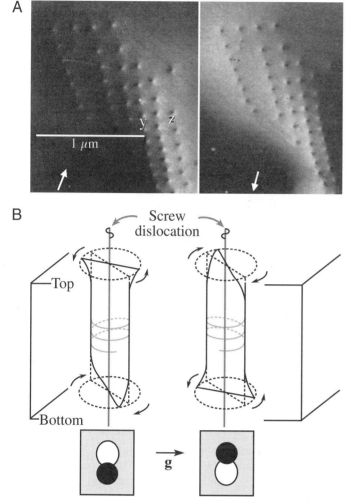

Figure 25.19. (A) Screw dislocations viewed end on (±**g**). The schematics (B) show the diffracting planes rotating in the same direction away from the edge-on orientation at both surfaces.

- Two grains may have a near-common plane and therefore a nearly common, but different, **g**-vector.
- In small-angle grain boundaries, θ is small, so the separation of the dislocations is large ($\sin \theta/2 \cong b/2d$).

The Σ = 3 twin boundary in fcc materials is an example of an interface where you can use common, but different, **g**-vectors. Here, the $(33\bar{3})$ plane is parallel to the $(\bar{5}11)$ plane, so these two **g**-vectors are identical, as you can see in Figure 25.22. However, this common reflection would not normally be used because **g** is rather large. This coincidence can also occur for other grain boundaries.

In the case of small-angle boundaries, we can pretend that the reflection is common to both grains, as illustrated in Figure 25.23. What we are really doing is treating

Figure 25.20. Dislocations interacting with a grain boundary; the dislocation contrast changes because its strain field changes when it enters the boundary and becomes part of the dislocation structure.

the dislocations as if they were isolated lattice defects; actually, the **g**-vectors for the two grains will be rotated relative to one another.

Lattice misfit is very important whenever we are studying thin films; dislocations are often present to accommodate the misfit. An example is shown in Figure 25.24, where dislocations are present between spinel and

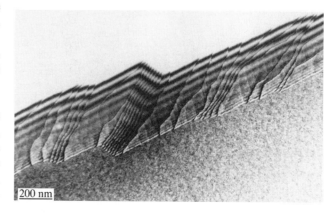

Figure 25.21. Steps at interfaces may also cause diffraction contrast. In this Ge specimen, the steps displace the thickness fringes in the GB. The fringe spacing is different at the top and bottom of the boundary because the diffraction conditions are different in each grain.

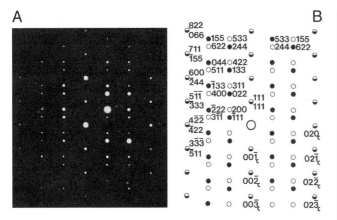

Figure 25.22. (A) DP and (B) its indexed schematic for a $\Sigma = 3$ twin boundary in an fcc material. Notice that many pairs of **g**-vectors exactly overlap but have very different indices.

NiO; these two materials both have the same fcc crystal structure. Although you can easily appreciate the change in lattice parameter, you must remember that there is also a less obvious change in the elastic constants.

> This means that the strain field at phase boundaries is *not* the same as at a grain boundary.

The TEM beam "sees" yet another change: the extinction distance is different. The result is that, if the crystal is inclined to the electron beam, you will see thickness fringes associated with the interface. Not much work has been done on this, but you may find that it is more difficult to

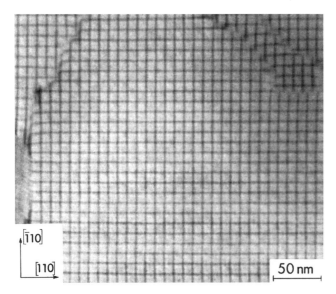

Figure 25.23. A low-angle (001) twist boundary in Si oriented almost exactly parallel to the specimen surface. Two (040) reflections were excited to form this BF image, but for small misorientations these are so close that we treat them as one reflection.

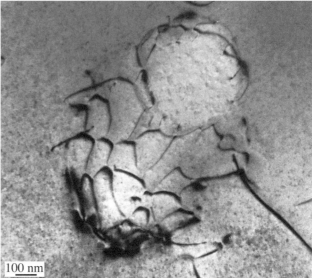

Figure 25.24. An irregular array of misfit dislocations at the interface between a spinel particle and an NiO matrix. The lattice mismatch is very small as you can appreciate from the scale. Although you can "see" a distorted hexagonal array of dislocations, you have to remember that this interface is actually curving within the specimen so that we are only seeing a projection of the structure.

use the **g·b** criterion for determining Burgers vectors, especially when the misfit is large.

Phase transformations often involve the movement of dislocations generally at semicoherent interfaces. All the conditions discussed above may hold; however, now the dislocations will certainly be associated with a step on the interface, so as to physically translate the interface as the transformation proceeds. However, you will find it dif-

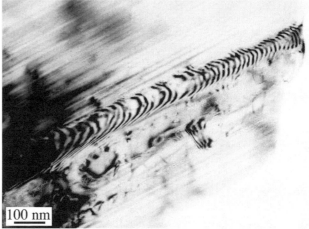

Figure 25.25. Transformation dislocations in the interface between a growing lath of hematite (pseudo-hexagonal alumina structure) in a ferrite (cubic spinel structure) matrix. The dislocations are curved because they were moving while heating the thinned specimen, which is why we know they are transformation dislocations, not simply misfit dislocations.

25 ■ STRAIN FIELDS

ficult to model the contrast from such dislocations, especially when you have a thin layer of the new phase enclosed by the matrix, as in the case when a precipitate grows, as illustrated in Figure 25.25.

The main effect of steps on such interfaces is that they cause a shift in the thickness fringes. It is often difficult to tell if there is also a dislocation present.

We'll summarize some features you should remember when studying dislocations in interfaces:

- If the orientation of the grains is different, the distribution of strain from the dislocation may be different in the two grains; the diffraction contrast is determined by this strain field.
- If the chemistry of the two grains is different or if you use different but equal **g**-vectors, the extinction distances will be different and the image of the dislocations must therefore be affected.
- Be careful not to confuse moiré fringes with dislocations (we'll discuss moiré fringes in Chapter 27). The guide is that the dark and light moiré fringes have approximately equal widths; if there is any ambiguity, you should use weak-beam imaging (Chapter 26) and carefully examine the DP.

Humble and Forwood (1975) have shown, using computer simulation of dislocations in interfaces, that it is best to use diffraction conditions where a reflection is satisfied in both grains, otherwise the dislocation images tend to be rather featureless relative to the interface thickness fringes.

25.10. VOLUME DEFECTS AND PARTICLES

When the defects are small, the image may be dominated by the strain-field contrast, and that is the aspect we are considering here. You have to remember, though, that the defects may have a different structure, lattice parameter, and composition. The theory for a spherical particle in a matrix was given over 30 years ago by Ashby and Brown (1963).

> The theory works well for coherent particles but as soon as the first interface dislocation appears, analysis becomes much more difficult.

Lattice-strain effects around spherical precipitates appear as lobes of low intensity with a line of no contrast perpendicular to **g**, as shown in Figure 25.26. If you measure the size of the precipitates from a DF image and the size of the strain-contrast lobes in BF, you can get a direct measure of

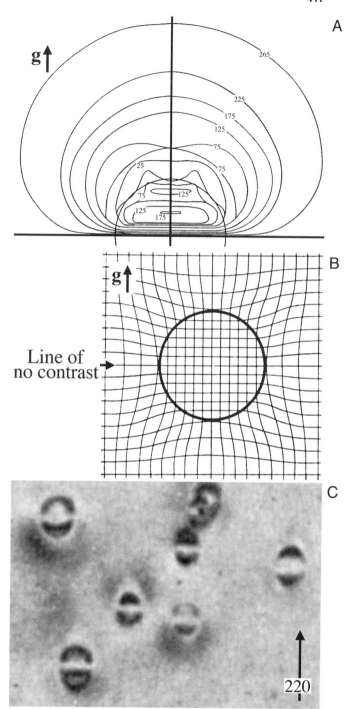

Figure 25.26. (A) Intensity contours from a simulated image of a particle like that shown schematically in (B). Notice the line of no contrast which corresponds to the plane that is not distorted by the strain field of the particle. (C) Experimental image of coherent particles in Cu-Co showing strain contrast.

the lattice strain surrounding a single precipitate, which is really quite remarkable. The process requires some specific experimental conditions and careful calibration of the photographic recording procedure, which there isn't space to describe, so you should read the original references for details. If your precipitates aren't spherical, intuitive interpretation of the images is unreliable and you have to resort to computer simulation.

Figure 25.26B shows how a spherical particle might strain the lattice. Notice that in this case, all the planes continue across the particle so it is coherent and there are no misfit dislocations. The figure here assumes that all the strain occurs in the matrix, which is only true for a hard particle in a soft matrix. The displacement field used to model this situation is

$$\mathbf{R} = C_\varepsilon \mathbf{r} \quad [25.14]$$

when $\mathbf{r} \leq \mathbf{r}_0$, and

$$\mathbf{R} = C_\varepsilon \frac{r_0^3}{r^3} \mathbf{r} \quad [25.15]$$

when $\mathbf{r} \geq \mathbf{r}_0$, where C_ε is an expression for the elastic constants, given by

$$C_\varepsilon = \frac{3K\delta}{3K + 2E(1+\nu)} \quad [25.16]$$

and K is the bulk modulus of the precipitate; E and ν are Young's modulus and Poisson's ratio, respectively, for the matrix. The important feature is that **R** always shows radial symmetry. Thus, when we consider the Howie–Whelan equations, we realize that when **g**·**R** = 0 we will see no contrast. So, there will be a "line of no contrast" normal to **g**.

The strain can be plotted using the equations given by Ashby and Brown and the image simulated (see below) as shown in Figure 25.26A. In the image from a specimen of a Cu-Co alloy containing small Co precipitates shown in Figure 25.26C, we can see that the images of the particles resemble butterflies or coffee beans. With the improvement in computers, the image contrast expected from much more complex particle geometries can now be calculated and can even consider statistical structural fluctuations (e.g., Karth *et al.* 1995).

25.11. SIMULATING IMAGES

It is important that you understand the origins of diffraction contrast from strain fields before you try to simulate this contrast using a computer. Having said that, few students would want to calculate image intensities by hand. The Howie–Whelan equations can be used to simulate images of dislocations, which is especially important when the dislocations are close together. The principal approaches used to simulate diffraction-contrast images were discussed in Sections 24.11 to 24.13.

> If you want to make quantitative comparisons with real images recorded on film, you must correct for the nonlinearity of the film (see Chapter 30).

Although the algorithm employed by the Head *et al.* (1973) programs allow very fast computation of the image, it does so by restricting the geometry of the defects. To cope with more general geometries, e.g., the strain field from end-on screw dislocations or nonparallel dislocations, Thölen (1970a) introduced a matrix algorithm. As we saw in Chapter 24, Howie and Basinski (1968) extended the two-beam calculations to include several beams on the systematic row and presented a method for circumventing the column approximation.

25.11.A. The Defect Geometry

When choosing the optimal simulation method, depending on the defect geometry, the problem of calculating the image belongs to one of three categories:

- *Two-dimensional problem:* including the most general geometries where integration of the full two-dimensional (*x*,*y*) grid is necessary.
- *One-dimensional problem:* geometries where the image depends only on either *x* or *y* and can be represented by a profile, e.g., problems involving a dislocation parallel to the foil surfaces.
- *GCS problem:* geometries where the method of generalized cross sections (GCS), developed by Head *et al.* (1973), can be applied. Situations where the dislocations and fault planes are parallel to each other, but inclined to the foil surface, are included in this group.

Choosing the best method can speed up the simulations considerably, as we'll show later. The Head *et al.* program automatically determines the category and selects the appropriate calculation method.

25.11.B. Crystal Defects and Calculating the Displacement Field

The program Comis can simulate amplitude contrast from any number of defects consisting of fault planes and straight, infinite dislocations (Rasmussen and Carter 1991). You just need to define the Burgers vector, line direction, and relative position; planar faults are defined by the plane normal, the displacement vector, and the relative position. You can predefine certain standard geometries to ease the process of defining the defect system.

Once you've defined the defect geometry, consider the region of the crystal you want to simulate. In situations where the "interesting" region is well defined (as in the case of inclined dislocations or intersecting dislocations), the program will determine this region and provide it as the default. However, you can always set the image region manually in Ångström units, to obtain a desired magnification.

The displacement field for the dislocations is calculated using linear, anisotropic elasticity theory (Eshelby *et al.* 1953) and is based on the algorithms given by Head *et al.* (1973), so you must specify the elastic constants of your crystal. The displacement field then corresponds to straight, infinite dislocations in an infinite medium with no account taken of surface relaxations. You can introduce image dislocations outside the crystal in order to include surface effects.

25.11.C. The Parameters

An example which shows simulated images of an orthogonal network of screw dislocations is given in Figure 25.27. Thölén (1970b) has analyzed this situation in detail. Comis can calculate the equilibrium configuration of certain types of interacting dislocations (Morton and Forwood 1973) using anisotropic elasticity theory, and then directly incorporate the resulting geometry in subsequent image simulations. As you can appreciate from equation 25.5, in such simulation studies, you will need *all* the parameters for the defects, the specimen, and the diffraction conditions:

- The foil thickness.
- The stacking-fault energy.
- The absorption parameters, usually using $|U'_g|/|U_g| = 0.1$.
- The number of beams included in the calculation.

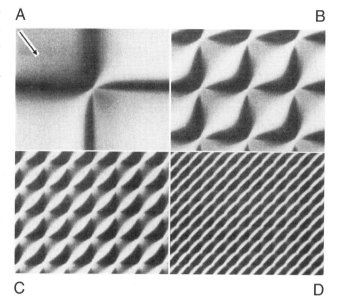

Figure 25.27. Simulated two-beam BF images of networks of screw dislocations, located in the middle of a foil with thickness equal to 4 times the extinction distance, ξ_g; $\mathbf{g}\cdot\mathbf{b} = 1$ for both dislocation types. The separation between the dislocations is: (A) ∞, (B) $1\xi_g$, (C) $0.5\xi_g$, (D) $0.25\xi_g$.

- The zone axis and the diffracting vectors.
- Also required are the electron energy, the elastic constants, the normal to the foil surface, the Burgers vectors, and the line direction of the dislocations.

The exact beam direction is then specified by defining the "center" of the Laue zone, giving the coordinates in terms of the **g** vector and $\mathbf{g}_z$. Here $\mathbf{g}_z$ is a specially defined vector in reciprocal space, which is automatically set to lie in the ZOLZ and to be perpendicular to **g**. Thus, if you place the center of the ZOLZ at (0, 0), the specimen is oriented on the zone axis; if you place the center of the ZOLZ at (0.5, 0), it corresponds to being at **g**/2, i.e., at the Bragg position with the **0** and **g** beams excited. If you change the second coordinate to give, say, (0.5, 0.5), you need to include beams from off the systematic row.

CHAPTER SUMMARY

The central idea of this chapter is that the strain field moves atoms off their perfect-crystal positions. We've concentrated on dislocations because the edge dislocation gives the clearest illustration of how the deformation produces the contrast and its structure can be understood with a two-dimensional projection. We can summarize the topics of the chapter as follows:

- There is a new feature to the column approximation. The displacement moves atoms out of the column and brings others into the column.

- The basis of the **g·b** analysis of dislocations is simply that the contrast is determined by **g·R(r)** and that **R(r)** is linearly related to **b**. For the screw dislocation, **R(r)** is directly proportional to **b**. For the edge dislocation, the image can also be affected by a **g·b×u** component which is caused by the buckling of the dislocation glide plane.
- Dislocation images are usually asymmetric. The contrast depends on the sign of (**g·b**)s.
- As a practical rule, we usually set **s** to be >0. Then the distortion due to the defect will bend the near-diffracting planes back into the Bragg-diffracting condition to give strong contrast. When **s** > 0, detail in the image is more localized relative to the defect than if we use the **s** = 0 condition.

There are many other situations which are closely related to the topics we've discussed in this chapter. For example, we have not discussed strain contrast associated with crack tips (de Graf and Clarke 1993) or the analysis of buckling of thin specimens (Thölén and Taftø 1993). Although these are rather specialized situations, they do illustrate the growing applications of diffraction contrast in the TEM.

REFERENCES

General References

Amelinckx, S. (1964) *The Direct Observation of Dislocations,* Academic Press, New York. A fascinating summary of the early studies by TEM.

Amelinckx, S. (1979) in *Dislocations in Solids,* 2 (Ed. F.R.N. Nabarro), North-Holland, New York. If you're interested in dislocations there are many other volumes in this set.

Amelinckx, S. and Van Dyck, D. (1992) in *Electron Diffraction Techniques,* 2 (Ed. J.M. Cowley), p. 1, Oxford University Press, New York.

Edington, J.W. (1976) *Practical Electron Microscopy in Materials Science,* Van Nostrand Reinhold, New York.

Head, A.K., Humble, P., Clarebrough, L.M., Morton, A.J., and Forwood, C.T. (1976) *Computed Electron Micrographs and Defect Identification,* North-Holland, New York.

Hirsch, P.B., Howie, A., Nicholson, R.B., Pashley, D.W., and Whelan, M.J. (1977) *Electron Microscopy of Thin Crystals,* 2nd edition, Krieger, Huntington, New York.

Hirth, J.P. and Lothe, J. (1982) *Theory of Dislocations,* 2nd edition, John Wiley & Sons, New York. The definitive textbook, but not for the beginner.

Hull, D. and Bacon, D.J. (1984) *Introduction to Dislocations,* 3rd edition, Pergamon Press, New York. A great introductory text.

Matthews, J.W., Ed. (1975) *Epitaxial Growth,* Parts A and B, Academic Press, New York.

Nabarro, F.R.N. (1987) *Theory of Dislocations,* Dover Publications, New York.

Porter, D.A. and Easterling, K.E. (1992) *Phase Transformations in Metals and Alloys,* 2nd edition, Chapman and Hall, New York.

Smallman, R.E. (1985) *Modern Physical Metallurgy,* 4th edition, Butterworth-Heinemann, Boston.

Sutton, A.P. and Balluffi, R.W. (1995) *Interfaces in Crystalline Materials,* Oxford University Press, New York.

Wolf, D. and Yip, S., Eds. (1992) *Materials Interfaces: Atomic-Level Structure and Properties,* Chapman and Hall, New York. A collection of review articles.

The original series of papers by P.B. Hirsch, A. Howie, M.J. Whelan, and H. Hashimoto in *Proc. Roy. Soc. London* **A252**, 499 (1960), **263**, 217 (1960), **267**, 206 (1962), and **268**, 80 (1962) are strongly recommended.

Specific References

Amelinckx, S. (1992) in *Electron Microscopy in Materials Science* (Eds. P.G. Merli and M.V. Antisari), World Scientific, River Edge, New Jersey.

Ashby, M.F. and Brown, L.M. (1963) *Phil. Mag.* **8**, 1083 and 1649.

de Graf, M. and Clarke, D.R. (1993) *Ultramicroscopy* **49**, 354.

Eshelby, J.D., Read, W.T., and Shockley, W. (1953) *Acta Metall.* **1**, 251.

Goringe, M.J. (1975) in *Electron Microscopy in Materials Science* (Eds. U. Valdrè and E. Ruedl), p. 555, Commission of the European Communities, Luxembourg.

Hirsch, P.B., Howie, A., and Whelan, M.J. (1960) *Phil. Trans. Roy. Soc.* **A252**, 499.

Howie, A. and Basinski, Z.S. (1968) *Phil. Mag.* **17**, 1039.

Howie, A. and Sworn, H. (1970) *Phil. Mag.* **31**, 861.

Hughes, D.A. and Hansen, N. (1995) *Scripta Met. Mater.* **33**, 315.

Humble, P. and Forwood, C.T. (1975) *Phil. Mag.* **31**, 1011 and 1025.

Karth, S., Krumhansl, J.A., Sethna, J.P., and Wickham, L.K. (1995) *Phys. Rev. B* **52**, 803.

Morton, A.J. and Forwood, C.T. (1973) *Cryst. Lattice Defects* **4**, 165.

Rasmussen, D.R. and Carter, C.B. (1991) *J. Electron Microsc. Technique* **18**, 429.

Steeds, J.W. (1973) *Anisotropic Elastic Theory of Dislocations,* Clarendon Press, Oxford, United Kingdom.

Takayanagi, K. (1988) *Surf. Sci.* **205**, 637.

Thölén, A.R. (1970a) *Phil. Mag.* **22**, 175–182.

Thölén, A.R. (1970b) *Phys. Stat. Sol. (a)* **2**, 537.

Thölén, A.R. and Taftø, J. (1993) *Ultramicroscopy* **48**, 27.

Tunstall, W.J., Hirsch, P.B., and Steeds, J.W. (1964) *Phil. Mag.* **9**, 99.

Wilkens, M. (1978) in *Diffraction and Imaging Techniques in Material Science,* 2nd edition (Eds. S. Amelinckx, R. Gevers, and J. Van Landuyt), p. 185, North-Holland, New York.

Weak-Beam Dark-Field Microscopy

26

- 26.1. Intensity in WBDF Images 423
- 26.2. Setting s_g Using the Kikuchi Pattern 423
- 26.3. How to Do WBDF 425
- 26.4. Thickness Fringes in Weak-Beam Images 426
- 26.5. Imaging Strain Fields 427
- 26.6. Predicting Dislocation Peak Positions 428
- 26.7. Phasor Diagrams 430
- 26.8. Weak-Beam Images of Dissociated Dislocations 432
- 26.9. Other Thoughts 436
  - 26.9.A. Thinking of Weak-Beam Diffraction as a Coupled Pendulum 436
  - 26.9.B. Bloch Waves 436
  - 26.9.C. If Other Reflections Are Present 437
  - 26.9.D. The Future 437

CHAPTER PREVIEW

The term "weak-beam microscopy" refers to the formation of a diffraction-contrast image in either BF or DF. The DF approach has been more widely used, in part because it can be understood using quite simple physical models. It also gives stronger contrast; we see white lines on a dark gray background. This chapter will be concerned only with the DF approach. Historically, the weak-beam dark-field (WBDF, often abbreviated to WB) method became important because, under certain special diffraction conditions, dislocations can be imaged as narrow lines which are approximately 1.5 nm wide. Equally important is the fact that the positions of these lines are well defined with respect to the dislocation cores; they are also relatively insensitive to both the foil thickness and the position of the dislocations in the specimen. The technique is particularly useful if you are studying dissociated dislocations where pairs of partial dislocations may only be ~4 nm apart and yet this separation greatly affects the properties of the material.

We first choose a particular **g** and bring this onto the optic axis as if intending to form a regular on-axis DF image. We then tilt the specimen to make s_g large and examine the DF image using reflection **g**. If a defect is present, the diffracting planes may be bent locally back into the Bragg-diffracting orientation to give more intensity in the DF image. The problem is that, as we increase s_g, the average intensity decreases as $1/s^2$; in the DP the beam appears as a weak spot, hence the name. When s_g is large, the coupling between **g** and the direct beam becomes small and the diffracted beam is said to be "kinematically diffracted." So, this chapter is where we will discuss the "kinematical approximation."

- You will sometimes see reference to the **g**(3**g**) WB condition. Beware! Sometimes you don't need to be this weak; sometimes this is not weak enough.
- It is often not the fact that **s** is large that is important; what is important is that ξ_{eff} is small.

This chapter is unusual in that it deals with a special imaging technique, rather than a concept or theory. Also, WBDF is only really useful when the specimen is not perfect, i.e., when you are interested in defects in the specimen or small changes in thickness. Therefore, you can skip this chapter if crystal-lattice defects are not relevant to your microscopy study. If you are interested in defects, you will find that this chapter really covers much more than WB microscopy. For example, we will use concepts developed for diffracted beams and carefully set the excitation error, s_g, by referring to the Kikuchi-line pattern. In Section 26.9 we will discuss some of the ways that new developments in TEM design are changing the way we do WB microscopy and how we interpret the images.

Weak-Beam Dark-Field Microscopy

26

26.1. INTENSITY IN WBDF IMAGES

We showed in Chapter 11 that, in a two-beam situation, the intensity of the diffracted beam **g** in a perfect crystal can be written as

$$\left|\phi_g\right|^2 = \left(\frac{\pi t}{\xi_g}\right)^2 \cdot \frac{\sin^2(\pi t s_{eff})}{(\pi t s_{eff})^2} \quad [26.1]$$

Remember that when we derived this expression we assumed that only two beams, O and G, are important. We will consider complications which arise when more beams are present in Section 26.9. The important variables in equation 26.1 are the thickness, t, and the effective excitation error, s_{eff}, which is given by

$$s_{eff} = \sqrt{s^2 + \frac{1}{\xi_g^2}} \quad [26.2]$$

In the WB technique we increase s to about 0.2 nm^{-1} so as to increase s_{eff}. (In most WB papers you will see this value as 2×10^{-2} Å^{-1}.) This large value of s means that s_{eff}, and therefore the intensity, I_g, become independent of ξ_g except as a scaling factor for t (in the prefactor in equation 26.1). The actual value of **s** can be set by carefully positioning the Kikuchi lines for the systematic row of reflections which includes **g**. You can best appreciate this effect by calculating a range of values for **s**. Remember when doing this that you must specify **g** and the kV because, as we saw in Chapter 11, ξ_g varies with both the reflection used to form the WB image and the energy of the electrons

$$\xi_{eff} = \frac{\xi_g}{\sqrt{w^2+1}} \quad [26.3]$$

A comment on equation 26.2: if $s \gg \xi_g^{-2}$ then $s \approx s_{eff}$, so that equation 26.2 becomes what is known as the "kinematical equation"; the kinematical equation cannot be applied for small **s** unless the thickness, t, is also very small.

Practical Considerations: As the value of s_{eff} increases, equation 26.1 shows that the intensity of the G beam decreases very rapidly. The result is that the exposure time needed to record the image on a photographic film also rapidly increases and has, in the past, been the factor which limited the usefulness of the technique. Although manufacturers may guarantee a drift rate of less than 0.5 nm per minute for new machines, values of six times this rate are common on many older instruments. The problem can be partly overcome by using photographic film with a fast emulsion or by modifying photographic processing conditions. In either case, the grainy appearance of the photographic emulsion would be increased. The problem of drift can, in principle, be overcome by using a video system to record the image and capturing frames from the video. You could then reduce the noise by frame averaging, particularly if you can take account of any drift. The causes of drift (specimen and thermal effects) and their correction or minimization are discussed in Chapter 8, but it is worth reminding you that change in the temperature of the water in the objective lens is a major cause of drift. Although WBDF imaging may aim for 0.5-nm resolution rather than 0.2 nm in HRTEM, exposure times are often 10 times greater for WBDF than HRTEM, so drift may be even more important.

26.2. SETTING s_g USING THE KIKUCHI PATTERN

Since the contrast in the WB image is so dependent on the value of s_g we need a method for determining s_g. We draw a line through the **g**-systematic row and let the Ewald sphere cut through this line at $n\mathbf{g}$, where n is not an integer. Figure 26.1 illustrates this situation. How can we "see" what the value of n is? Of course we can't, since we are looking ap-

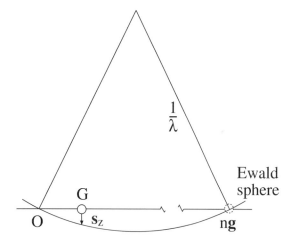

Figure 26.1. The Ewald sphere construction showing the diffraction conditions used to obtain weak-beam images. The sphere cuts the row of systematic reflections at "$n\mathbf{g}$" where n is not necessarily an integer.

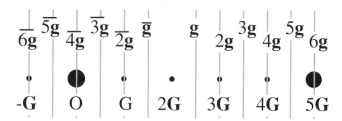

Figure 26.3. A schematic diagram showing the positions of the Kikuchi lines for the systematic row of reflections when 4**g** is excited.

proximately normal to the ZOLZ. We can't just judge n by looking at the intensities of the spots except in special circumstances.

The solution to the problem can be appreciated by looking at the Kikuchi pattern shown in Figure 26.2. When **g** is exactly at the Bragg condition, the **g** Kikuchi line passes through the **g** reflection; when 3**g** is exactly satisfied, the 3**g** Kikuchi line passes through the 3**g** reflection. We might guess that n is ~3.2 in Figure 26.2, but we don't have a 3.2**g** Kikuchi line; we have to deduce this value of n from the position of the 3**g** Kikuchi line. Remember (from Chapter 19) that when the 3**g** Kikuchi line passes through 3.5**g** on the **g**-systematic row, the 4**g** Kikuchi line and the Ewald sphere pass through the 4**g** reflection, as shown in Figure 26.3. Therefore, when the Ewald sphere passes through 3.2**g**, the 3**g** Kikuchi line will pass through 3.1**g**; we can express this simple geometric result as

$$n = 2m - N \quad [26.4]$$

where $N\mathbf{g}$ refers to the Kikuchi line closest to $n\mathbf{g}$ (N is an integer) and $m\mathbf{g}$ is the location of the Kikuchi line as we measure it. In the example above, we can choose N to be 3 so that, if m is 3.1, then n is 3.2; if, instead, we choose N to be 4, then m is 3.6 (because we measure the position of the 4**g** Kikuchi line) and n is still 3.2. Having determined n we need to estimate **s**. This we do using the expression

$$\mathbf{s} = \tfrac{1}{2}(n-1)|\mathbf{g}|^2 \lambda \quad [26.5]$$

which you can derive from Figure 26.4 using the intersecting chord theorem ($ab = cd$) and the fact that $1/\lambda$ is much larger than **s**.

You can immediately appreciate some important results from this expression:

- Setting $n = -1$ gives the same value of **s** as for $n = 3$, but the sign is reversed.
- The magnitude of **s** is more strongly dependent on $|\mathbf{g}|$ than on λ, but it depends on *both*.
- The specific nature of the material enters through **g**, the microscope affects **s** through λ. Here we recommend that you use a spreadsheet to calculate different values of **s** as you vary **g** or λ. A selection of these for Cu and Si is given in Table 26.1.

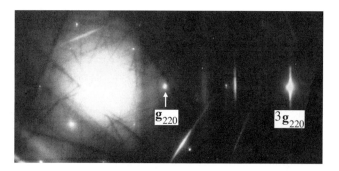

Figure 26.2. A DP obtained when the specimen is tilted to a suitable orientation for WB microscopy. Here **g** is a 220 reflection and 3**g** is strong.

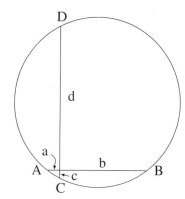

Figure 26.4. The intersecting chord construction used to deduce the value of $\mathbf{s}_g$: we approximate d to $2/\lambda$, c is **s**, a is $|\mathbf{g}|$, and b is $(n-1)|\mathbf{g}|$.

Table 26.1. Values of S (Å⁻¹) for Various kV[a]

Si	Cu	Accelerating Voltage			
n_{Si}	n_{Cu}	100 kV	200 kV	300 kV	400 kV
4.9	2.8	0.020	0.014	0.011	0.009
6.9	3.6	0.030	0.020	0.016	0.013
8.5	4.3	0.038	0.025	0.020	0.017
9.9	4.9	0.044	0.030	0.024	0.020

[a]For Cu, a = 0.3607 nm; for Si, a = 0.534 nm. The values of n are accurate to better than 0.1. In both cases the calculation is for **g** = 220.

As an exercise, you can use equations 26.2 and 26.5 to calculate s_{eff} with ξ_g = 42 nm for the 220 reflection in Cu at 100 kV. You can then see when s_{eff} becomes "independent" of ξ_g. Next, repeat the exercise with other values of λ or for other reflections and materials.

One point you should bear in mind is that none of the above discussion requires a particular value for **s**, and yet you will often read that **s** must be ≥ 0.2 nm⁻¹ for a WB image. This value of **s** is recommended when you are studying defects quantitatively, because computer calculations show that the position of the image can then be directly related to the position of the defect. You will find that smaller values of **s** will often give you WB images which contain the information you want and you can more easily see and record the image!

26.3. HOW TO DO WBDF

The nature of the WB image imposes a restriction on the maximum specimen thickness which you can use because the visibility of such images decreases as the thickness increases (due to a corresponding increase in inelastic scattering). However, the orientation of the specimen is accurately set by reference to the Kikuchi lines observed in the DP (Chapter 19), and these are not visible in specimens which are too thin. Therefore, your specimen thickness must be greater than a certain minimum value. Also, if the observations you made on certain defects, in particular dislocation ribbons and nodes, are to be interpreted as representative of the bulk and not influenced by the surface of the foil, then again your foil must not be too thin. You can generally satisfy these requirements by selecting defects for detailed study which, in the case of Cu and its alloys, lie in areas which are about 70 nm thick.

Due to the very low intensity of the WB images, exposure times required are typically on the order of 4 to 30 seconds using Kodak SO-163 film. The main factor limiting the exposure time is the inherent instability of any specimen stage. To minimize the exposure time, you can usually use a highly convergent (or divergent) beam, contrary to the assumption implicit in the simple theory. The effect of this convergence is that oscillations in image intensity and position, which result from variations in depth and thickness parameters, are diminished.

Step-by-step: We'll go through how to set up the **g**(3.1**g**) diffraction mode condition since this is widely used in practice. Actually, we generally refer to it as "**g**(3**g**) with s_{3g} positive" because we guess the value of n by estimating m. This condition ensures that the 3**g** reflection is not satisfied and also that you can use the BF 0(**g**) image, with s_g slightly positive, to locate the defect and to focus your image. The first two steps are illustrated in Figure 26.5, relating what happens in the Ewald sphere model with what you see happening to the Kikuchi lines.

- Orient the specimen in BF so that **g** is excited and s_g is just greater than zero. Make sure that no other reflections are excited.
- Use the DF beam-deflecting coils to bring the reflection **g** on to the optic axis. Use the binoculars because **g** becomes very weak; underfocus the beam before you use the high-resolution screen.
- Insert the objective aperture. In BF, check that the aperture is centered, then switch to DF and check that the spot G is centered in the aperture.
- Fine-tune your conditions, looking at the DP with G centered.
- Go to imaging mode; you now have a WB image with the required **g**(3.1**g**) condition.

Since you have inserted a small objective aperture, you should now check the objective astigmatism. We use a small objective aperture so as to remove inelastic scattering; remember that this aperture will then limit our potential resolution. If you focus the beam, you may change the position of the beam and probably the astigmatism! Remove the objective aperture and check that no other reflections are strongly excited when you are in DF. Then repeat the process starting at the third step (insert the objective aperture).

After finely focusing the image, record it together with its SAD pattern.

If you're not sure why this "trick" for setting the **g**(3.1**g**) condition works, go back to Chapter 19 and draw the systematic row and the corresponding Kikuchi lines. Then move the spots, while keeping the crystal, and thus the Kikuchi lines, fixed. You may see the diffraction condition **g**($\bar{g}$) used. This was the original condition suggested by Cockayne *et al.* (1969); it does give you the same values of **s** as the **g**(3.1**g**) condition, but the interband scattering processes are different and it is not so convenient to change from BF to WBDF. However, you may find variations on

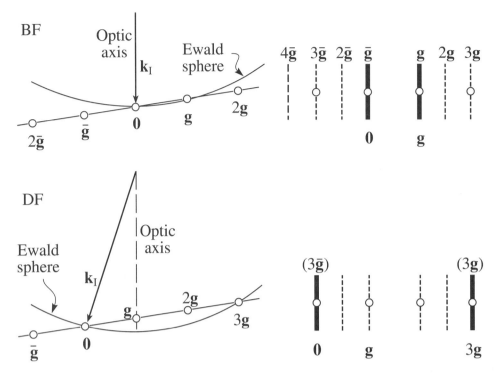

Figure 26.5. Relationship between the orientation of the Ewald sphere and the position of the Kikuchi lines for the **0**(**g**) (upper) and **g**(3**g**) (lower) diffraction conditions. The two pairs of diagrams are related by tilting the beam; the specimen has not tilted so the position of the Kikuchi lines is unchanged.

this condition useful when you need to use a **g**(n**g**) condition with a large value of n.

Weak-beam microscopy becomes much easier when you are comfortable with using the TV camera. Just being able to see whether the image is moving due to specimen drift can save boxes of photographic film. As we mentioned earlier, you could use frame averaging to reduce the noise. However, you will realize that the extra magnification from the video system tends to limit the area which can be viewed, so that film is still preferred for most images. You will find that 30 k× magnification on the plate is a good compromise; without video, you will use 50-60 k× to allow you to focus with the binoculars.

26.4. THICKNESS FRINGES IN WEAK-BEAM IMAGES

Thickness fringes in WB images are just like thickness fringes in strong-beam images but the effective extinction distance, ξ_{eff}, is much smaller. From equation 26.1 we can see that the intensity minimum occurs at thicknesses of $\mathcal{N}(s_{eff}^{-1})$ with maxima at $(\mathcal{N} + \frac{1}{2})(s_{eff}^{-1})$. The effective extinction distance for $s = 0.2$ nm^{-1} is 5 nm; this value is rather sensitive to the precise value of s so that the fringes will change if the foil bends. Using WB images we can form a rather detailed contour map of the specimen, but you must remember that both surfaces may be inclined to the beam, as shown in Figure 26.6.

The thickness effect is illustrated in Figure 26.7. These images were recorded with $s = 0.2$ nm^{-1}. The MgO specimen has been heat-treated so that there are large regions where the surface is atomically flat on both sides.

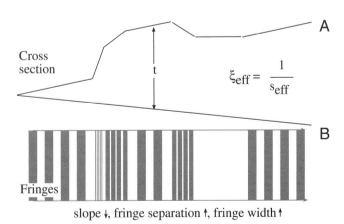

Figure 26.6. (A) In WB imaging, the thickness periodicity depends on the effective extinction distance, ξ_{eff}. (B) The separation of the fringes varies accordingly.

Figure 26.7. (A) WB thickness fringes from annealed MgO. (B,C) Higher magnification of regions B and C. Compare to Figure 23.3.

26.5. IMAGING STRAIN FIELDS

The principle of the technique is very simple. When the area of your specimen in which the defect of interest lies is oriented away from the Bragg position, the reflecting planes may be bent back into the reflecting position close to the defect. The region over which this occurs is very small because the strain has to be quite large to cause this bending. For the (220) planes in Cu (which fixes the plane spacing, d), the planes must rotate through an angle of ~2° to change s locally from 0.2 nm^{-1} to zero.

> The intensity of the reflection that we see in the DP is still small even though a relatively intense peak may occur in the image close to the defect core because the DP averages over a large area.

When you look at dislocations in the WB image, you see bright lines on a dark background. Let's compare the WB image to a BF image of the same defect in Figure 26.8. You can see that the WBDF image is much narrower;

Before heating, the specimen had been acid etched, which caused the holes seen in this image (they're black because we're in DF). The specimen shows inclined steps which curve across the surface. Where we see wide uniform gray regions the surface is atomically flat. At A there is a large inclined step which runs into the hole B. Notice how the number of fringes around the hole increases at A. Around the holes (Figures 26.7B and C) we see much more closely spaced fringes because the thickness changes more quickly here. Now if we look at the edge of any hole such as C, we see that the spacing of the fringes has one value away from the hole but another, smaller value close to the hole. What we find is that the inclined surface facets on different planes with each facet becoming steeper closer to the hole. This topology is a result of the way the specimen has been prepared and would not normally be found, say, in electro-chemically polished specimens, but it does illustrate the possibility of "profiling" using thickness fringes.

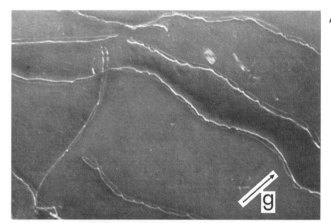
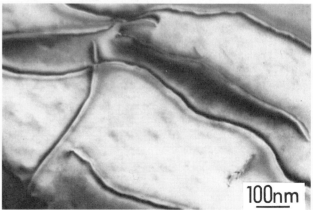

Figure 26.8. A comparison of dislocation images in a Cu alloy formed using (A) WB and (B) strong-beam ($s_g > 0$) conditions.

you could make the comparison look even better if you make **s** very close to zero in the BF image.

We'll keep our discussion of dislocations brief but draw your attention to a few particular points:

- In the WB technique, most of the specimen is tilted so that **s** is large; the lattice planes in most of the specimen are then rotated away from the Bragg condition. However, as you can see in Figure 26.9, near the core of the dislocation the planes are locally bent back into the Bragg condition.
- This bending is only large close to the core of the dislocation (i.e., at the same depth from the surface).
- The peak you see in the WB image is always displaced to one side of the dislocation core. If you reverse the sign of **g**, the peak moves to the other side of the core. If you reverse the Burgers vector, **b** (rotate the diagram in Figure 26.9 through 180°), but keep **g** the same, the peak again moves to the other side.

- If you increase **s** in the crystal, then the planes must bend more to satisfy the Bragg condition, which means the peak will move closer to the dislocation core.
- When we say "position of the peak," we are always talking about a projected position where the projection is along $\mathbf{k}_D$.
- There will be some situations where the strain is not large enough to compensate for the **s** you have chosen. Then you will only see poor contrast in the image.

26.6. PREDICTING DISLOCATION PEAK POSITIONS

There are three ways to calculate the contrast in a WB image. Since each teaches us something new, we'll go through them in turn.

Method 1: The WB Criterion states that the largest value of ϕ_g in the WB image occurs when s_R, which we derived in equation 25.7, is zero. We can express this result as

$$s_R = s_g + \mathbf{g} \cdot \frac{d\mathbf{R}}{dz} = 0 \qquad [26.6]$$

Equation 26.6 tells us that if the effective value of s (i.e., s_{eff}) is zero, even though s_g is not zero, then the direct beam and the diffracted beam, **g**, are strongly coupled. In this situation the strain field effectively rotates the lattice planes into the Bragg-reflecting position (Hirsch *et al.* 1977). Therefore, the crystal can be oriented so that ϕ_g is small for all columns except those near the dislocation core, where it can attain a considerable magnitude due to the strong coupling with the transmitted beam as it passes through the region close to the core of the dislocation where s_{eff} is zero. This increased amplitude is then retained below the core when the coupling between the two beams is decreased again. The intensity is expected to be largest for that column where s_{eff} remains closest to zero over the longest length, and this occurs for the column where there is an inflection in the curve of **R** versus z (Cockayne *et al.* 1969, Cockayne 1972 and 1981). Therefore, the position of the WB peak should occur when equation 26.6 is satisfied at a turning point of $\mathbf{g} \cdot (d\mathbf{R}/dz)$.

Method 2: The Kinematical Integral. An alternative criterion for defining the position of the WB peak was derived by Cockayne (1972). In the approximation where only two beams are considered and **s** is sufficiently large, Cockayne showed that the maximum scattering from the transmitted to the diffracted beam occurs where the kinematical integral, defined as

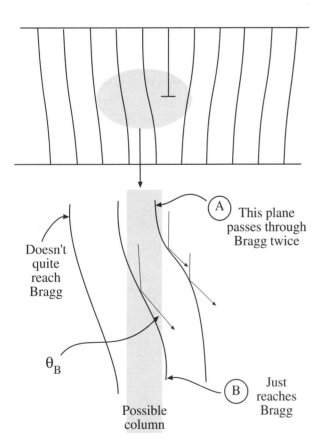

Figure 26.9. WB images of defects show high intensity close to the defect because only there are the diffracting planes bent back into the Bragg condition. This illustration is for an edge dislocation.

$$\int_{column} e^{\left\{-2\pi i\left(s_g z + \mathbf{g}\cdot\mathbf{R}\right)\right\}} dz \qquad [26.7]$$

is maximized. This maximum, in general, occurs for a column which is closer to the dislocation core than predicted by *Method 1*. The reason for this difference is interesting: because the planes are bent, the reciprocal lattice point is, on average, nearer to the Ewald sphere. Therefore, the integral has a larger value over the length of the column.

Without doing all the math, we can illustrate how these two approaches are related. What we want to do is determine when ϕ_g is large, but still kinematical (i.e., **s** is large); we want to maximize the kinematical integral in equation 26.7. We can do this using the stationary-phase method described by Stobbs (1975). We write the integral as

$$\int_0^t \exp\left(-2\pi i\left[\frac{z^2}{2}\cdot\frac{d^2}{dz^2}(\mathbf{g}\cdot\mathbf{R}) - \frac{z^3}{3}\cdot\frac{d^3}{dz^3}(\mathbf{g}\cdot\mathbf{R})\right]\right) dz \qquad [26.8]$$

where we have set $s + d/dz\,(\mathbf{g}\cdot\mathbf{R}) = 0$. If we also set $d^2/dz^2(\mathbf{g}\cdot\mathbf{R}) = 0$ (at the inflection), we go further to ensuring that the term in the square brackets is zero. This condition is what we guessed for the first method of defining the WB criterion.

Method 3: Compute the Contrast. Now that personal computers are widely available, we can calculate the position of the WB peak and graph the results. What we then find is that the WB peak actually lies between the two values predicted by the two criteria deduced using the first two methods. We also find, using the computer, that the position and width of the image peak are affected by any strongly excited diffracted beams so these *must* be avoided. A practical point is that the computer sometimes gives a rather pessimistic view of the variability in the peak position, so we have to weight the results carefully. Remember that the important advantages of this approach are that we can include the effects of the other diffracted beams which are always present, and we can take account of other variables, such as the convergence of the beam.

In the kinematical approximation, the half-width, Δx, of the image of an undissociated screw dislocation with $|\mathbf{g}\cdot\mathbf{b}| = 2$ is given approximately by the relation derived by Hirsch *et al.* (1960)

$$\Delta x = \frac{1}{\pi\, s_{eff}} \simeq \frac{\xi_{eff}}{3} \qquad [26.9]$$

This expression is a very useful rule of thumb. You will realize that this WB image width is special for three reasons, which arise because it doesn't depend on ξ_g. So, once **s** is fixed in a WB image, we can make several surprising statements regarding the width of the dislocation peak:

- It does not depend on the material.
- It does not depend on the reflection.
- It does not depend on the kV.

Take the example of the 220 reflection in Cu at 100 kV: ξ_g is 42 nm, and the width Δx is 14 nm. So, even if equation 26.9 is slightly wrong, the image width is greatly reduced in WB. If we make $s_g = 0.2$ nm^{-1}, then the half-width is 1.7 nm.

Computed many-beam images confirm that dislocations in other orientations give rise to similar narrow peaks when this value of s_g is used. A series of peak profiles for different values of t is shown in Figure 26.10. Notice that although the intensity of the peak may be only

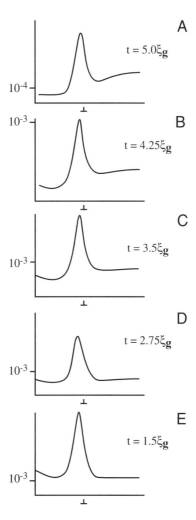

Figure 26.10. Examples of computer-calculated intensity peaks in WB images of an edge dislocation in Cu for different values of t. The intensity is relative to unit incident beam intensity.

about 0.1% of the incident beam, it is still much higher than the background.

> For WB microscopy, $s_g = 0.2$ nm^{-1} is always a useful guide; it satisfies the requirements that the image should have a narrow width and show a high contrast between the defect and background regions.

Equation 26.9 indicates that as we increase the value of s, the half-width of the image peak decreases. However, a maximum is imposed on s by the fact that the intensity of the diffracted beam varies as s^{-2}. If we make s much larger, the contrast of the image therefore becomes too small to be of practical use.

The basic requirements governing the value of s which you must use for quantitative imaging are:

- $s \geq 2 \times 10^{-2}$ Å^{-1} to give sufficiently narrow peaks for fine detail to be studied.
- $s \leq 3 \times 10^{-2}$ Å^{-1} because the intensity varies as s^{-2} in the kinematical limit.
- $s\xi_g \geq 5$ to give sufficient contrast in the WB image.

If you use the **g**(3**g**) condition for Cu with **g** = 220 and 100-keV electrons, then the value of s_g will be 0.238 nm^{-1}.

26.7. PHASOR DIAGRAMS

We sometimes find it useful to demonstrate the depth dependence of the contrast in the WB image using phasor, or amplitude-phase, diagrams which we introduced in Chapter 2. You can generally use such diagrams whenever the kinematical approximation holds; they are equivalent to a graphical integration of the two-beam equations for the case where **s** is large. In fact, many of the early calculations of defect contrast were made using this approach before computers became widely available. We recommend that you glance at the original paper by Hirsch *et al.* (1960).

The basic idea is shown in Figure 26.11. We simply add all the $d\phi_g$ increments to ϕ_g. In doing so, we take account of the phase changes which occur as the beam passes through the crystal. Remember that in this approximation, no electrons leave the **g** beam! If the crystal is perfect and our increments are sufficiently small, we will produce a smooth circle.

The circumference of this circle is ξ_{eff} as we require for the depth periodicity, and the radius is $\xi_{eff}/2\pi$ or $(2\pi s_{eff})^{-1}$. Notice that as we increase **s**, we decrease ξ_{eff} and the circle

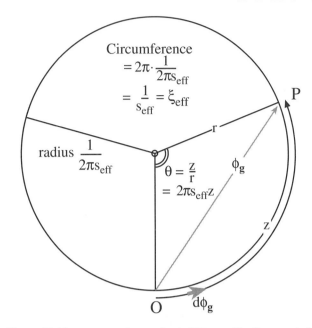

Figure 26.11. A phasor diagram for the WB case. The distance z is the arc OP measured around the circumference and the radius of the circle is $(2\pi s_{eff})^{-1}$.

becomes smaller. Thus we move around the circle more quickly if **s** is large. In other words, our effective extinction distance is reduced, as we knew from Chapter 13.

If the diffracted beam passes through a stacking fault, it will experience an extra phase shift given by $2\pi \mathbf{g} \cdot \mathbf{R}$. Using the familiar example for an fcc crystal, we take an example where $\mathbf{R} = \frac{1}{3}[11\bar{1}]$ and $\mathbf{g} = (20\bar{2})$, which gives $\alpha = 2\pi/3 = 120°$ (modulo 2π). The abrupt phase change is shown at P_3 in Figure 26.12. Now, the value of ϕ_g (P_1P_2) can be much larger than in a perfect crystal. The locus of ϕ_g still travels round the first circle until it meets the

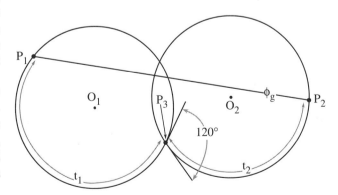

Figure 26.12. A phasor diagram used to explain the WB contrast from a SF at depth t_1 in a foil of thickness $t_1 + t_2$. P_1 is the top and P_2 the bottom of the foil. Here, the phase change at the SF (P_3) is 120°.

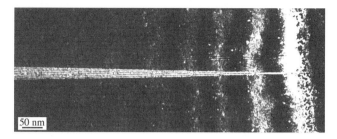

Figure 26.13. Illustration of thickness fringes in an experimental image of two inclined SFs in a wedge specimen of spinel. The thickness fringes from the wedge and from the inclined SF can be clearly seen and counted. The second SF shows little contrast because it has a different **R** value.

planar fault at $z = n\xi_{eff} + t_1$, where z and t_1 are measured parallel to $\mathbf{k}_D$. It then moves onto the second circle until it reaches $z = t$. You can readily see that if we keep the depth of the fault, t_1, fixed, we then see depth fringes which vary with periodicity ξ_{eff} as we vary the total thickness, t. (The value of α is still 120°.) The situation is a little more difficult to envision if we keep t fixed but vary t_1, but the principle is the same. Images of WB fringes at inclined SFs are illustrated in Figure 26.13. The thickness fringes of the wedge specimen and those from the inclined stacking fault can both be clearly seen and counted. Notice that the number of bright fringes on the planar defect really does increase by one for every increment ξ_{eff} in the thickness of the wedge and you don't need to know **R**.

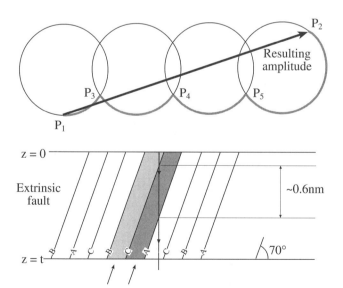

Figure 26.14. Phasor diagram for a series of overlapping planar defects at P_3, P_4, and P_5. For a 111 foil normal, the inclined $11\bar{1}$ planes lie at 70° to the surface so that the spacing between adjacent planes in the direction of the beam is 0.627 nm. Compare with Figure 26.12.

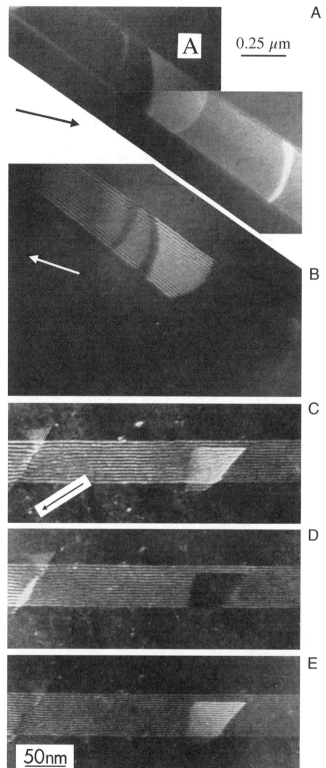

Figure 26.15. (A,B) Overlapping SFs imaged in WBDF conditions for ±**g**. Many more fringes occur than in the BF image in Figure 24.5. Fringes occur at A where the BF image showed no contrast. (C–E) Changes in fringe spacing and intensity in the overlapping region as **s** increases.

You can imagine applying this analysis to the situation where you have several overlapping planar faults, as shown in Figure 26.14. ϕ_g (P_1P_2) can become very large (does our approximation of unit incident intensity still hold?). This situation does occur in practice, as shown in Figure 26.15. Here we have several overlapping faults. The bright ones are bright even when ϕ_g reaches its minimum, as you would expect from Figure 26.14. If you compare the WB image with its strong-beam counterpart in Figure 24.5, you will notice that there is much more detail in the WB image; as we saw in Chapter 24, overlapping faults on nearly adjacent planes can essentially give no contrast in the BF image. You can easily check such effects in WB by adjusting **s** as in Figures 26.15C–E. It's interesting to realize that this effect can occur even when two intrinsic SFs lie on adjacent planes to give the extrinsic SF (Föll *et al.* 1980, Wilson and Cockayne 1985). This approach can be used to image other planar defects, such as the {112} twin boundary which can have a significant thickness (Carter *et al.* 1995). The key factor is that, under WB conditions, ξ_{eff} can become comparable to the distance that the beam travels between encountering successive planes of atoms, particularly when the interface is quite steeply inclined relative to the beam.

We can also use a phasor diagram to describe the contrast from a dislocation, but now the phase change occurs over a range of thickness rather than at a particular value. As illustrated in Figure 26.16, the phase can either add or subtract depending on the sign of **g**·**R**. When the phase changes quickly with a change in *t*, as at the center of Figure 26.16, it means that we are strongly coupling the incident and diffracted beams.

We'll summarize our discussion of phasor diagrams with two points:

- They should only be used when the kinematical approximation holds.
- They then give us a graphical method for understanding the variation of ϕ_g with thickness, especially when crystal defects are present.

26.8. WEAK-BEAM IMAGES OF DISSOCIATED DISLOCATIONS

Although the study of dislocations is a very specialized topic, it beautifully illustrates the potential of the WB technique. Dissociated dislocations are common in face-centered-cubic (fcc) materials (including Si) and ordered intermetallics such as Ni_3Al. The geometry of a dissociated dislocation in Cu is summarized in Figure 26.17. We gave some general references on the theory of dislocations in Chapter 25.

Since the computed many-beam images show that the position of the dislocation image lies close to the position predicted by the WB criterion, this criterion is used in practice because it allows us to deduce an equation for the position of the image. We can thus directly relate the separation of Shockley partials, for example, to the measured separation of the two peaks observed in $|\mathbf{g}\cdot\mathbf{b}_T| = 2$ images of dissociated dislocations. Hence we can estimate the stacking-fault energy (SFE) of semiconductors and several fcc metals. (See Carter 1984 for a review of dissociated dislocations and Geerthsen and Carter 1993 for a comparison of WB and HRTEM.) In order to interpret WB images of extended dislocation configurations, we need to know how the position of the image peak is related to the position of the dislocation core. This information is essential whenever we use the WB technique to collect *quantitative*

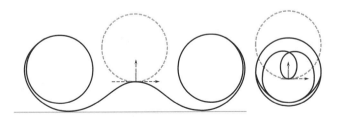

Figure 26.16. Phasor diagrams for a dislocation for ±**g**. The phase change is not abrupt but rather occurs over an extended distance along the column. In the left diagram the phase change is the direction shown in Figure 26.12, but in the right diagram the phase change has the opposite sign.

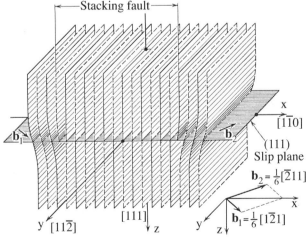

Figure 26.17. The geometry of a perfect dislocation in Cu. The perfect dislocation separates into two Shockley partial dislocations with Burgers vectors $\mathbf{b}_1$ and $\mathbf{b}_2$ separated by a SF on a {111} plane.

data. We will now answer two questions for quantitative WB analysis:

- What factors determine the position of the WB peak?
- What methods are used to determine the Burgers vectors of the dislocations?

When the WB conditions are satisfied, dissociated dislocations can be imaged in a {220} reflection for which $|\mathbf{g}\cdot\mathbf{b}_T| = 2$, so that each partial dislocation gives rise to an intensity peak having a half-width of 1 to 1.5 nm. The separation of the partials can be deduced to within ±0.7 nm providing that the peak separation is greater than ~2.5 nm. For many materials, we need to use anisotropic elasticity theory to relate the atomic displacements to the image, so the computer becomes essential.

Why not just use HRTEM since, as we'll see in Chapter 28, HRTEM can give detail down to below 0.2 nm, whereas WB is often limited to ~1 nm? If you want to interpret an HRTEM image the defect must be *absolutely straight*, parallel to the beam, and located in a very thin region of the specimen. The segment of dislocation studied by HRTEM will thus be no longer than 20 nm and less than 10 nm for the highest resolution. In a WB image the defect can be micrometers long and, in a relatively thick foil, it can even change direction. If you look back at Figure 26.8 you'll see pairs of lines which correspond to partial dislocations in the WB image. You can see other features, such as constrictions. The strong-beam image may also show two or more lines for a particular dislocation but, as we saw in Chapter 25, these lines are not related to the detailed structure of the dislocation but rather to n in the equation $\mathbf{g}\cdot\mathbf{b} = n$.

Ideally, for quantitative analysis, you should choose long, nearly straight dislocations. As shown in Figure 26.17, the two Shockley partial dislocations lie in the (111) plane of the foil. The Burgers vectors of the total and partial dislocations can be determined by imaging the dislocation in the WB mode using the {2$\bar{2}$0} reflections. A sharp peak is found for a partial dislocation with $|\mathbf{g}\cdot\mathbf{b}_p| = 1$, and either no peak or a diffuse one (arising from the anisotropy of the lattice) if the partial dislocation has $|\mathbf{g}\cdot\mathbf{b}_p| = 0$. When $|\mathbf{g}\cdot\mathbf{b}_T| = 2$, the diffraction vector $\mathbf{g}$ and the Burgers vector $\mathbf{b}_T$ are parallel and, in the image of a dissociated dislocation, two sharp peaks are formed, one corresponding to each of the partial dislocations (both now have $|\mathbf{g}\cdot\mathbf{b}_p| = 1$).

> One of the peaks is on average more intense than the other; the order reverses when $\bar{\mathbf{g}}(3\bar{\mathbf{g}})$ is used instead of $\mathbf{g}(3\mathbf{g})$.

We can explain this difference in intensity if one peak (the weaker) arises from the region between the partial dislocations and the other from outside the dissociated dislocation. This effect cannot occur in $|\mathbf{g}\cdot\mathbf{b}_T| = 1$ images, when $|\mathbf{g}\cdot\mathbf{b}_p| = 1$ for one partial and $|\mathbf{g}\cdot\mathbf{b}_p| = 0$ for the other, and it can be used to identify the reflection for which $|\mathbf{g}\cdot\mathbf{b}| = 2$. Confirmation of the Burgers vector is always obtained using the BF mode, observing characteristic $|\mathbf{g}\cdot\mathbf{b}_T| = 2$ or $|\mathbf{g}\cdot\mathbf{b}_T| = 0$ images.

In a WB image with $|\mathbf{g}\cdot\mathbf{b}_T| = 2$, each of the partial dislocations will generally give rise to a peak in the image which is close to the dislocation core (Cockayne *et al.* 1969). You can calculate the approximate positions of these peaks using the criterion from equation 26.6. Then, you can relate the separation of the peaks in the image to the separation Δ of the partial dislocations.

We can write the displacement, using isotropic elasticity theory, as the sum of the displacements due to the individual partial dislocations. If the Burgers vector of a straight, mixed dislocation lies in the (111) plane parallel to the surface of a foil, then at a distance x from the dislocation core

$$-s_g = \frac{|\mathbf{g}|}{2\pi}\left[\left(|\mathbf{b}_1| + \frac{|\mathbf{b}_{1e}|}{2(1-v)}\right)\frac{1}{x} + \left(|\mathbf{b}_2| + \frac{|\mathbf{b}_{2e}|}{2(1-v)}\right)\frac{1}{x-\Delta}\right] \quad [26.10]$$

Here x defines an axis perpendicular to both the dislocation line and the beam direction, and e refers to the edge component of the Burgers vectors of the partial dislocations, 1 and 2. This relation is particularly simple because, for the geometry we have chosen, the term $\mathbf{g}\cdot\mathbf{b}\times\mathbf{u}$ is zero. Using the notation

$$a = -s_g\left[\frac{|\mathbf{g}|}{2\pi}\left(|\mathbf{b}_1| + \frac{|\mathbf{b}_{1e}|}{2(1-v)}\right)\right]^{-1} \quad [26.11]$$

and

$$b = -s_g\left[\frac{|\mathbf{g}|}{2\pi}\left(|\mathbf{b}_2| + \frac{|\mathbf{b}_{2e}|}{2(1-v)}\right)\right]^{-1} \quad [26.12]$$

equation 26.10 reduces to

$$1 = \frac{1}{ax} + \frac{1}{b(x-\Delta)} \quad [26.13]$$

which has two solutions, x_+ and x_-, given by

$$x_\pm = \frac{ab\Delta + a + b \pm \left[(ab\Delta + a + b)^2 - 4ab^2\Delta\right]^{\frac{1}{2}}}{2ab} \quad [26.14]$$

These values of x define the positions of the peaks in the image. The separation between these peaks is then given by

$$\Delta_{obs} = \left[\Delta^2 + \frac{(a+b)^2}{a^2b^2} + \frac{2(a+b)\Delta}{ab} - \frac{4\Delta}{a}\right]^{\frac{1}{2}} \quad [26.15]$$

We can rearrange this equation to make it appear more symmetric in a and b. Of course, it will not be symmetric because the peak is always located on one side of the dislocation.

$$\Delta_{obs} = \left[\left(\Delta + \frac{1}{b} - \frac{1}{a}\right)^2 + \frac{4}{ab}\right]^{\frac{1}{2}} \quad [26.16]$$

Computed images confirm that this relation is accurate for $\Delta_{obs} > 2.5$ nm, to within ± 0.7 nm (Cockayne 1972). This uncertainty is due to the variation of the peak position with the depth of the dislocation in the foil and the foil thickness. A small uncertainty arises when you have not determined the actual direction of $\mathbf{b}_T$, i.e., whether it is in the direction of $\mathbf{g}$ or $\bar{\mathbf{g}}$. Stobbs and Sworn (1971) have found, using anisotropic elasticity theory, that the relation (equation 26.16) subject to the ± 0.7 nm uncertainty is still a good approximation.

As a simple exercise, consider the WB images of a dissociated screw dislocation and a dissociated edge dislocation. You should pay particular attention to "a" and "b," because in one case $\mathbf{b}_{1e}$ and $\mathbf{b}_{2e}$ have the same sign, while in the other the sign is opposite. Does the image always have the same width when you reverse $\mathbf{g}$?

Example 1. Even if you never want to calculate the actual separation of two dislocations from observations of two peaks, you can learn new ideas about dislocations from such images. Figure 26.18 is a famous set of images showing a dislocation in Si which is constricted along part of its length and dissociated along the rest. Even if you don't know the precise details of the dislocation structure, you know that it can adopt two variants; the rest of the task is modeling the defect.

Example 2. The WB image of the node pair in Figure 26.19A tells you very quickly that the two nodes are different; if we form images using other $\mathbf{g}$ vectors, (B–D)

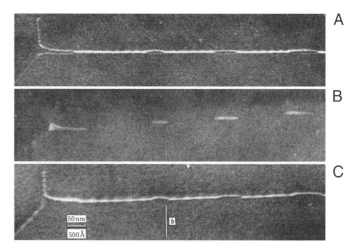

Figure 26.18. WB image of a dislocation in Si which has both dissociated and constricted segments: (A) $\mathbf{g}\cdot\mathbf{b} = 2$; both partial dislocations are visible. (B) $\mathbf{g}\cdot\mathbf{b}_T = 0$ showing SF contrast. (C) $\mathbf{g}\cdot\mathbf{b} = 1$; only one partial dislocation is visible.

we find that one of the partial dislocations is out of contrast in the image. The extended node contains the same type of intrinsic stacking fault that is present in the dissociated dislocation; $\mathbf{g}\cdot\mathbf{R}$ is zero in this image for the stacking fault. The other node is constricted, within the detectability of the WB technique. Comparison with the BF images in Figure 25.8 is instructive.

Example 3. We mentioned that the peak moves to the other side of the dislocation if we reverse $\mathbf{b}$. This is exactly what happens for a dislocation dipole as you can see in Figure 26.20. This is a complicated figure except that you can use it not only to see the inside–outside contrast in WB images of dislocation dipoles but also as an exercise in $\mathbf{g}\cdot\mathbf{b}$ analysis. A dislocation dipole is a pair of dislocations identical in every way, apart from the sign of the Burgers vector. If we now reverse $\mathbf{g}$, then both peaks move to the other side of their respective dislocations. This change in contrast is referred to as inside–outside contrast and is commonly seen on dislocation loops, which are themselves closely related to these dipoles (just more "equiaxed"). The images shown in Figures 26.20B,C illustrate the dramatic change in contrast which you can see on reversing $\mathbf{g}$. Some of the dipoles completely disappear in (A–C) because they are a special form of defect known as a faulted dipole. Such dipoles usually give very low contrast in strong-beam BF images because the dislocations are always very close together so that their strain fields overlap, and the lattice is thus only distorted over very small distances. When we use the WB technique, we are probing the structure on these very small dimensions and the contrast can be high. Again, compare with the BF image in Figure 25.13.

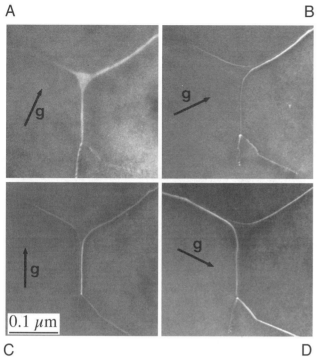

Figure 26.19. WB images of a pair of dislocation nodes formed by the dissociation of interacting dislocations lying on a {111} plane in a Cu alloy. The SF is imaged in (A) and, imaging with different reflections (B–D), the partial dislocations are out of contrast when $\mathbf{g} \cdot \mathbf{b} = 0$. Compare with the BF image in Figure 25.8 and note the difference in magnification.

Examples 4 and 5. The WB technique allows us to see features which would be hidden if we used strong-beam imaging. Figure 26.21 gives an example where an inclined SF is cutting through several dissociated dislocations. The interaction of the two defects would be masked by the SF fringes in the strong-beam image but is clearly visible in this WB image. Seeing small particles close to dislocations is difficult in strong-beam imaging. Although not easy in WB, Figure 26.22 does illustrate that it can be done. These images show, for example, that the behavior of the dissociated dislocation is different on each side of the particle (see also Figure 26.23), and again stress the advantage of WB over BF imaging.

Example 6. We noted in Section 25.8 that the surface of the specimen can affect the geometry of the defects we are examining. In general, the specimen needs to be thinner for WB imaging than for strong-beam imaging. Therefore, surface effects can be even more important. Figure 26.23 shows an example where this effect is particularly clear. Dislocations which were uniformly separated in the bulk material now appear wedge-shaped: the effects of the two surfaces are different in this case (Hazzledine *et al.* 1975).

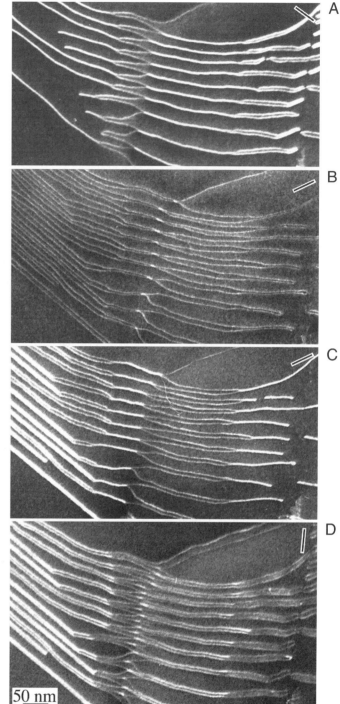

Figure 26.20. Four WBDF images showing an array of dislocation dipoles in a Cu alloy having a low stacking-fault energy. The reflections are all 220-type and the dislocations all lie on (111) planes which are nearly parallel to the surface of the specimen. All the dislocations are dissociated. All the Shockley partial dislocations are in contrast in (D) while half are out of contrast in (A–C). Notice that the narrower images are brighter than the wider ones: the strain is large in between the dislocations but decreases rapidly outside the dipole since the total Burgers vector of a dipole is zero.

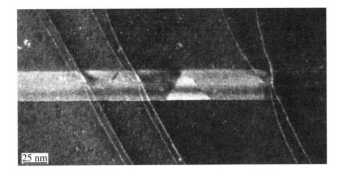

Figure 26.21. WB image of an inclined SF cutting through a series of dissociated dislocations lying parallel to the surface of the Cu-alloy specimen.

26.9. OTHER THOUGHTS

26.9.A. Thinking of Weak-Beam Diffraction as a Coupled Pendulum

We can illustrate the principle which underlies the increase in intensity in the WB image close to a dislocation using the mechanical analog of a coupled pendulum. A diagram is shown in Figure 26.24. The two pendula are connected (coupled) by a third string. If we start the left pendulum swinging but hold the connecting string, the right-hand pendulum remains stationary. Now release the connecting string. You will see that the right-hand pendulum now begins to swing. If we let the process continue, eventually the right-hand pendulum is swinging as much as the original one did, but the original one is stationary: this is the strong-beam analog! All the kinetic energy has been transferred from one pendulum to the other. Given more time, we will achieve the original condition. Now repeat the exercise, but hold the connecting string again after the right-hand pendulum has begun to swing; you will notice that both pendula continue to swing, each with a constant amplitude. The role of the connecting string is to couple the two pendula (beams) so that we transfer energy from one beam to the other. In WB TEM, the defect acts as the connecting string. The two beams are only coupled over a short length as they travel past the defect. We can plot this amplitude (or intensity); try this as an exercise.

26.9.B. Bloch Waves

We discussed Bloch waves in Chapter 14. The difficulty in applying Bloch-wave analysis to the WB situation is that we are usually interested in defects. However, we can make some basic comments. For the reflection **g** to give a WB image, $|\phi_g|$ must be much smaller than unity in the regions of perfect crystal but, in strained regions, a change

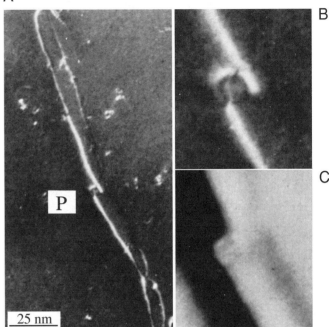

Figure 26.22. (A) WB image of a dissociated dislocation interacting with a particle (P) in a Cu alloy. (B) and (C) are enlargements of the WB and corresponding BF images, respectively.

$\Delta\psi^{(j)}$ in the amplitude of the Bloch wave j can give rise to a change $\Delta\phi_g$. Cockayne has shown that the appreciable contrast which can then be present in the WB image is due, in the two-beam approximation, to the interband scattering from Bloch wave 1 to Bloch wave 2. In the general case, the scattering is from the branch of the dispersion surface with the largest $\psi^{(j)}$ to those branches with the largest $C_g^{(j)}$, i.e., from the Bloch wave with the largest amplitude to the one which is most strongly excited. The dispersion surface for the **g**(3**g**) diffraction geometry is shown in Figure 26.25. It's an instructive exercise to reread this paragraph thinking how each statement relates to this figure, and to consider other diffraction geometries, e.g., **0**(2**g**).

Figure 26.23. WB image showing the dissociation of a group of dislocations which are inclined to the foil surface to give wedge-shaped SFs. The shape of the SFs is caused by surface stresses.

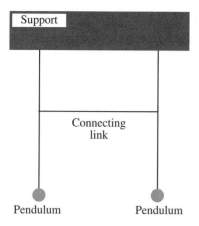

Figure 26.24. The coupled pendulum.

The Bloch-wave analysis of the problem leads to two further points which simplify the interpretation of WB images:

- The diffraction conditions should be such that only one interband scattering process is important.
- In the two-beam approximation, in order for the image peaks to show sufficient contrast, it is generally found that $w\ (=s\xi_g)$ is greater than ~5.

You can satisfy the first requirement by ensuring that no reflections are strongly excited. The second condition is usually already satisfied because of the more stringent requirement that s should be greater than 0.2 nm^{-1}. For example, for a $\{2\bar{2}0\}$ reflection in copper with 100-keV electrons, w is automatically greater than 8 since ξ_g is 42 nm.

The $\mathbf{g}(3.1\mathbf{g})$ condition may also be preferable to $\mathbf{g}(\bar{\mathbf{g}})$ on theoretical grounds if images are to be compared with computed profiles made using the column approximation, i.e., it simplifies your interpretation. The basis for this suggestion is that the region of the dispersion surface from which the scattering occurs is flatter for the $\mathbf{g}(3.1\mathbf{g})$ diffraction geometry than for the $\mathbf{g}(\bar{\mathbf{g}})$ case.

26.9.C. If Other Reflections Are Present

Several times in the previous discussion, we have said that no reflections should be strongly excited. Much of our thinking has been based on the two-beam approximation we introduced in Chapter 11. When you are using WB conditions, you must be even more careful. Consider the $\mathbf{g},3\mathbf{g}$ geometry shown in Figure 26.5. We form the WB image using reflection $\mathbf{g}$ so electrons are weakly scattered from the O beam into the G beam. However, once in the G beam, they can be strongly scattered into the 2G beam. We can picture this process by drawing the new Ewald sphere for the "new" incident beam, G; this sphere passes through 2G!

For the mathematically inclined, you can go back to the many-beam equations which we introduced briefly in Chapter 13. The coupling of beams $\mathbf{g}$ and $\mathbf{h}$ is determined by $(s_g - s_h)^{-1}$ and has an extinction distance given by ξ_{g-h}. If s_g and s_h are equal, then the coupling between these beams will be strong. Furthermore, the characteristic length for the coupling in this example will be ξ_{2g-g} or ξ_g, which is what you would have guessed from Figure 26.25.

26.9.D. The Future

Several new developments will change how we practice the WB technique. The main point here is that you should remember the principles because they will not change.

- Slow-scan CCD cameras give a very linear response and therefore make quantitative analysis of WB images possible. For this to happen, computer modeling of the defect and simulation of the image will be needed. We will return to this topic in Chapter 30.
- An FEG and energy-filtered imaging will allow us either to minimize the effect of variations in the energy or to form WB images using particular sections of the energy-loss spectrum. Then, we will need to extend the theory.
- Image processing and frame averaging should allow us to reduce the noise and again aid quantification.

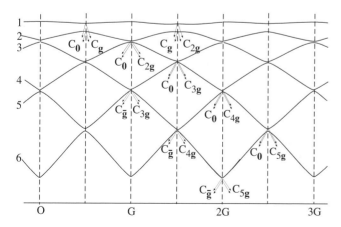

Figure 26.25. Dispersion surface construction which is used to describe the $\mathbf{g}(3\mathbf{g})$ geometry. The BZB is at 1.5G and tells you which reflections are strongly coupled; see Figure 15.9.

Image simulation, as we described in Chapter 25, will allow us to be more quantitative in our interpretation of WB images.

CHAPTER SUMMARY

The basic idea of the WB technique is very simple: using a large value of **s** gives a small ξ_{eff} and hence a narrow image of most defects, since the width of a dislocation is related to $\xi_{eff}/3$. What you should remember is that the value of *s* for a particular diffraction condition **g**(*n***g**) depends not only on *n* and **g**, but also the lattice parameter of the crystal and the wavelength of the electrons. You will see the "magic number" **s** = 0.2 nm^{-1} quoted often.

- Remember that this number gives a rule of thumb if you want to do quantitative analysis. It does not usually correspond to **g**(3**g**).
- Don't use the **g**(3**g**) condition without calculating the value of s_g.
- The term **s·R** has been neglected in this analysis; we usually assume that the deformable-ion model from Section 24.13 is valid.

You can often get all the information you need with less effort using a somewhat smaller value of **s**. As always, the longer you take to perfect the image, the more likely you are to alter your specimen, especially the defect structure.

Finally, remember that the diffracted beam travels parallel to k_D. Therefore, the image of any defect is also projected in this direction. Even though the Bragg angle is small, this means, for example, that the apparent separation of defects in the image may not be equal to their horizontal separation relative to their glide planes if the defects are located at different heights in the specimen. This projection error can vary, depending on the **g** and **s** used to form the image and the orientation of the specimen.

REFERENCES

Specific References

Carter, C.B. (1984) in *Dislocations–1984* (Eds. P. Veyssière, L. Kubin, and J. Castaing), p. 227, Editions du CNRS, Paris.

Carter, C.B., Mills, M.J., Medlin, D.L., and Angelo, J.E. (1995) 7th International Conference on Intergranular and Interphase Boundaries in Materials, Lisbon, Portugal.

Cockayne, D.J.H. (1972) *Z. für Naturforschung.* **27a**, 452.

Cockayne, D.J.H. (1981) *Ann. Rev. Mater. Sci.* **11**, 75.

Cockayne, D.J.H., Ray, I.L.F., and Whelan, M.J. (1969) *Phil. Mag.* **20**, 1265.

Föll, H., Carter, C.B., and Wilkens, M. (1980) *Phys. stat. sol. (A)* **58**, 393.

Geerthsen, D. and Carter, C.B. (1993) *Phys. stat. sol. (A)* **136**, 29.

Hazzledine, P.M., Karnthaler, H.P., and Wintner, E. (1975) *Phil. Mag.* **32**, 81.

Hirsch, P.B., Howie, A., Nicholson, R.B., Pashley, D.W., and Whelan, M.J. (1977) *Electron Microscopy of Thin Crystals*, 2nd edition, p. 164, Krieger, Huntington, New York.

Hirsch, P.B., Howie, A., and Whelan, M.J. (1960) *Proc. Roy. Soc. London* **A252**, 499.

Stobbs, W.M. (1975) in *Electron Microscopy in Materials Science,* Vol. **II** (Eds. U. Valdrè and E. Ruedl), p. 593, CEC, Brussels.

Stobbs, W.M. and Sworn, C.H. (1971) *Phil. Mag.* **24**, 1365.

Wilson, A.R. and Cockayne, D.J.H. (1985) *Phil. Mag.* **A51**, 341.

Phase-Contrast Images

27

- 27.1. Introduction 441
- 27.2. The Origin of Lattice Fringes 441
- 27.3. Some Practical Aspects of Lattice Fringes 442
  - 27.3.A. If s = 0 442
  - 27.3.B. If s ≠ 0 442
- 27.4. On-Axis Lattice-Fringe Imaging 442
- 27.5. Moiré Patterns 444
  - 27.5.A. Translational Moiré Fringes 445
  - 27.5.B. Rotational Moiré Fringes 445
  - 27.5.C. General Moiré Fringes 445
- 27.6. Experimental Observations of Moiré Fringes 445
  - 27.6.A. Translational Moiré Patterns 446
  - 27.6.B. Rotational Moiré Fringes 447
  - 27.6.C. Dislocations and Moiré Fringes 447
  - 27.6.D. Complex Moiré Fringes 448
- 27.7. Fresnel Contrast 450
  - 27.7.A. The Fresnel Biprism 450
  - 27.7.B. Magnetic-Domain Walls 451
- 27.8. Fresnel Contrast from Voids or Gas Bubbles 451
- 27.9. Fresnel Contrast from Lattice Defects 452
  - 27.9.A. Grain Boundaries 453
  - 27.9.B. End-On Dislocations 453

CHAPTER PREVIEW

We see phase contrast any time we have more than one beam contributing to the image. In fact, whenever we say "fringes," we are essentially referring to a phase-contrast phenomenon. Although we often distinguish phase and diffraction contrast, this distinction is generally artificial. For example, as we saw in Chapters 23 and 24, even thickness fringes and stacking-fault fringes are phase-contrast images although we usually think of them as two-beam diffraction-contrast images.

Phase-contrast imaging is often thought to be synonymous with high-resolution TEM. In fact, phase contrast appears in most TEM images even at relatively low magnifications. We will draw your attention to its role in the formation of moiré patterns and Fresnel contrast at defects. This Fresnel contrast has the same origin as that which we used in Chapter 9 to correct the astigmatism of the objective lens.

As with many of the topics we've discussed, we can approach the problem at several different levels. One danger is that you may be tempted to use one of the prepackaged simulation programs to predict phase-contrast images, without considering the limitations of such packages. We will begin this chapter by discussing some simple approaches to understanding phase-contrast effects as they relate to lattice-fringe imaging.

Phase-Contrast Images

27.1. INTRODUCTION

Contrast in TEM images can arise due to the differences in the phase of the electron waves scattered through a thin specimen. This contrast mechanism can be difficult to interpret because it is very sensitive to many factors: the appearance of the image varies with small changes in the thickness, orientation, or scattering factor of the specimen, and variations in the focus or astigmatism of the objective lens. However, its sensitivity is the reason phase contrast can be exploited to image the *atomic structure* of thin specimens. Of course, this also requires a TEM with sufficient resolution to detect contrast variations at atomic dimensions, and the proper control of instrument parameters that affect the phases of the electrons passing through the specimen and the lenses. If you know what you are doing, the procedures can be straightforward; the level of operator skill necessary to obtain such images can be acquired with practice by most TEM users.

The most obvious distinction between phase-contrast imaging and other forms of TEM imaging is the number of beams collected by the objective aperture or an electron detector. As described in the previous chapters, a BF or DF image requires that we select a single beam using the objective aperture. A phase-contrast image requires the selection of *more than one* beam. In general, the more beams collected, the higher the resolution of the image. However, we will see that there are reasons why some beams, which are apparently admitted through the aperture, might not contribute to the image. The details of this process depend on the performance of the electron-optical system. We'll first examine the theory and then consider the practical aspects.

27.2. THE ORIGIN OF LATTICE FRINGES

We can understand the origin of lattice fringes by extending the analysis of Chapter 13 to allow the two beams, **0** and **g**, to interfere, i.e., use the objective aperture to select only two beams. We begin by rewriting equation 13.5

$$\psi = \varphi_0(z) \exp 2\pi i (\mathbf{k}_I \cdot \mathbf{r}) + \varphi_g(z) \exp (2\pi i \mathbf{k}_D \cdot \mathbf{r}) \quad [27.1]$$

where we know

$$\mathbf{k}_D = \mathbf{k}_I + \mathbf{g} + \mathbf{s}_g = \mathbf{k}_I + \mathbf{g}' \quad [27.2]$$

We are thus using a two-beam approximation but allowing $\mathbf{s}_g$ to be nonzero. Now we will make some simple substitutions, setting $\varphi_0(z) = A$ and take $e^{2\pi i \mathbf{k}_I \cdot \mathbf{r}}$ out as a factor. We will also represent the expression for φ_g from equation 13.5 as

$$\varphi_g = B \exp i \delta \quad [27.3]$$

where

$$B = \frac{\pi}{\xi_g} \frac{\sin \pi t s_{eff}}{\pi s_{eff}} \quad [27.4]$$

and

$$\delta = \frac{\pi}{2} - \pi t s_{eff} \quad [27.5]$$

The $\pi/2$ in the expression for δ takes care of i in equation 13.5 and we'll pretend that the specimen is so thin that we can replace s_{eff} with s. Thus equation 27.1 becomes

$$\psi = \exp (2\pi i \mathbf{k}_I \cdot \mathbf{r})[A + B \exp i(2\pi \mathbf{g}' \cdot \mathbf{r} + \delta)] \quad [27.6]$$

The intensity can then be expressed as

$$I = A^2 + B^2 + AB[\exp i(2\pi \mathbf{g}' \cdot \mathbf{r} + \delta) + \exp -i(2\pi \mathbf{g}' \cdot \mathbf{r} + \delta)] \quad [27.7]$$

$$I = A^2 + B^2 + 2AB \cos\left(2\pi \mathbf{g}' \cdot \mathbf{r} + \delta\right) \qquad [27.8]$$

Now $\mathbf{g}'$ is effectively perpendicular to the beam so we'll set it parallel to x and replace δ, giving

$$I = A^2 + B^2 - 2AB \sin\left(2\pi \mathbf{g}'x - \pi st\right) \qquad [27.9]$$

Therefore, the intensity is a sinusoidal oscillation normal to $\mathbf{g}'$, with a periodicity that depends on s and t (just like thickness fringes). (Note that g and s are not bold in equation 27.9 because they represent the magnitude of the vectors, not the vectors themselves.) We can, with care, relate these fringes to the spacing of the lattice planes normal to $\mathbf{g}'$. Although we have obtained this equation using a very simple model, it gives us some useful insight, which will also be helpful when we talk about many-beam images in Chapter 28.

> The intensity varies sinusoidally with different periodicities for different values of g'.

This model will be equally valid even if the incident beam is tilted slightly off the optic axis.

27.3. SOME PRACTICAL ASPECTS OF LATTICE FRINGES

27.3.A. If s = 0

If we just have **0** and **g** in the objective aperture and we then set $\mathbf{s} = 0$ for reflection G (so $\mathbf{g}' = \mathbf{g}$), we will see fringes in the image (Figure 27.1A) which have a periodicity in the x direction of $1/g$, i.e., the spacing of the planes which give rise to **g**. This result holds wherever $\mathbf{s} = 0$, no matter how **0** and **g** are located relative to the optic axis, even if the diffracting planes are not parallel to the optic axis.

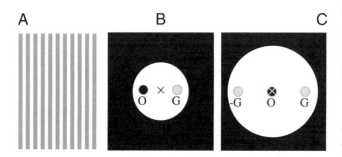

Figure 27.1. (A) Schematic tilted-beam 111 lattice fringes in Si formed using the O and G beams symmetrically displaced relative to the optic axis; **g** is normal to the fringes. (B) Ideal diffraction geometry to produce tilted-beam fringes. (C) On-axis three-beam geometry.

Figure 27.1B shows the ideal geometry for producing images like Figure 27.1A. It is called the "tilted-beam condition" and it means that the planes of interest lie parallel to the optic axis. If we use the geometry shown in Figure 27.1B we have $\mathbf{s} = 0$ and the planes are parallel to the optic axis but not parallel to the incident beam. Therefore, the fringes cannot correspond directly to the individual planes. If we use the on-axis geometry shown in Figure 27.1C, the planes are viewed edge on, but $\mathbf{s} \neq 0$ for reflection G; so we must also consider reflection –G.

27.3.B. If s ≠ 0

If the specimen is not exactly flat, then **s** will vary across the image; even if you set $\mathbf{s} = 0$ in the DP, it will not be zero everywhere. If **s** is not zero, then the fringes will shift by an amount which depends on both the magnitude of **s** and the value of t, but the periodicity will not change noticeably. We expect this **s** dependence to affect the image when the foil bends slightly, as is often the case for thin specimens. We also expect to see thickness variations in many-beam images, since ideally **s** is not zero for any of the beams; **s** may also vary from beam to beam.

27.4. ON-AXIS LATTICE-FRINGE IMAGING

We've just seen that two beams can interfere to give an image with a periodicity related to $|\Delta \mathbf{g}|^{-1}$. Since one beam is the direct beam, $|\Delta \mathbf{g}|^{-1}$ is just d, the interplanar spacing corresponding to **g**. If you align your beam parallel to a low-index zone axis, then you'll see fringes running in different directions; these fringes in the image must correspond to an array of spots in the DP. The spacings of the spots may be inversely related to the lattice spacings, as shown in Figure 27.2, which extends Figure 27.1 to the many-beam case. In general, this array of spots bears no direct relationship to the *position* of atoms in the crystal.

> The trouble is that the fringes look so like atomic planes that you can be easily misled into thinking that they are atomic planes.

We'll see more on this when we discuss image simulation in Chapter 29. In case you are in doubt, compare the beautiful image shown in Figure 27.3A with the projected structure of Si in Figure 27.3B. The Si dumbbells are a pair of atoms which are 1.4 Å apart in this projection of the structure. The aperture used to form the image included 13 reflections, as shown in Figure 27.3C. The diffi-

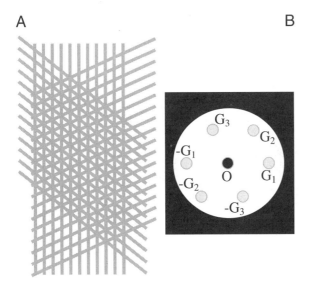

Figure 27.2. (A) Schematic many-beam image showing crossing lattice fringes and (B) the diffraction pattern.

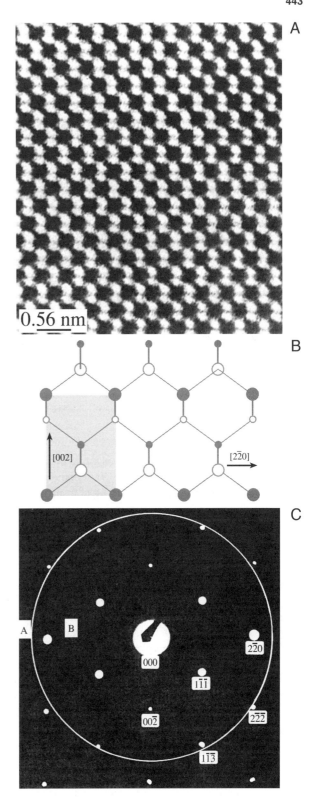

culty is that, in the image, the spots in the dumbbell are really just 1.3 Å apart but the point resolution of the TEM was only ~2.5 Å. You can see from the structure that the real dumbbell spacing corresponds to the (004) plane spacing, but the 004 reflection was *not* used to form the image. The explanation was given by Krivanek and Rez (1980); the dumbbells in the image are caused by the crossing {113} fringes.

The lesson is: we only knew the image did not correspond to the structure because we knew the structure! Taking this example as your guide, consider the case where a defect is present in an image where the perfect-crystal spots are all in the "correct" position. Could you still be certain that the details in the image close to the defect give you a true picture of the location of the atoms close to the defect? The answer is of course "no."

So lattice fringes are *not* direct images of the structure, but just give you lattice spacing information.

On-axis lattice-fringe images are perhaps best used as a measure of the local crystal structure and orientation. The exception, as we'll see in the next chapter, is when these images can only be interpreted using extensive computer simulation. Figure 27.4 illustrates some typical applications of the phase-contrast imaging mode, where we can learn a lot about our material by intuitive interpretation without the need for simulating their images.

Figure 27.4A shows interfaces between a spinel particle and an olivine matrix; Figure 27.4B shows how we

Figure 27.3. (A) On-axis image of a perfect Si crystal; (B) the projected structure; (C) the diffraction pattern showing the 13 spots used to form the image inside the aperture (ring). The Si dumbbells do not correspond to the closely spaced pairs of spots in the image.

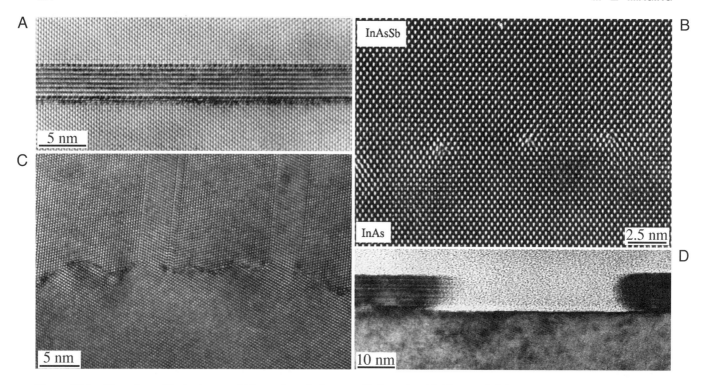

Figure 27.4. Illustrations of lattice images which contain easily interpreted information. (A) The spinel/olivine interface; (B) dislocations at a heterojunction between InAsSb and InAs; (C) a grain boundary in Ge faceting on an atomic scale; (D) a profile view of a faceted surface.

can locate dislocations at a heterojunction; Figure 27.4C shows the atomic-scale faceting of a grain boundary in Ge, and Figure 27.4D illustrates the faceting of a surface.

27.5. MOIRÉ PATTERNS

Moiré (pronounced "mwa-ray") patterns can be formed by interfering two sets of lines which have nearly common periodicities. We can demonstrate two fundamentally different types of interference: the rotational moiré and the translational (often referred to as misfit) moiré. It's easy to understand moirés if you make three transparent sheets of parallel lines (two with the same spacing and one slightly different): you can generate such sets of lines readily using any computer, choosing the line widths to be similar to the gaps between them. Then try these three exercises:

- Take two misfit sets and align them exactly. This gives you a set of moiré fringes which are parallel to the lines forming them, as shown in Figure 27.5A.
- Take two identical sets of lines and rotate them. Now, you produce a set of moiré fringes which is perpendicular to the average direction of the initial lines (Figure 27.5B).
- Take the first two sets and rotate them so you produce moiré fringes, as in Figure 27.5C; but note that their alignment to your reference sets is not obvious.

When the misfit or misorientation is small, the moiré-fringe spacing is clearly much coarser than that of the

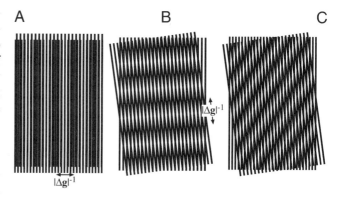

Figure 27.5. (A) Translational moiré fringes; (B) rotational moiré fringes; (C) mixed moiré fringes; note the relationship between the fringes and their constituent lattices.

27.5.A. Translational Moiré Fringes

lines themselves. In particular, if the sets of lines in Figure 27.5 are actually lattice planes in a crystal, the moiré fringes may give information about the crystals even if you cannot resolve the lattice planes. The simplest way to analyze the spacings and orientation of the moiré fringes is to consider the diffraction vectors from the two "lattices." Incidentally, the term "moiré" originated in the textile industry; it's related to the French word for "mohair," the silky hair of the Angora goat; hence the watery or wavy pattern seen in silk fabrics.

27.5.A. Translational Moiré Fringes

In this case, since the planes are parallel, the **g**-vectors will also be parallel. If we write these as $\mathbf{g}_1$ and $\mathbf{g}_2$, we produce a new spacing $\mathbf{g}_{tm}$ given by

$$\mathbf{g}_{tm} = \mathbf{g}_2 - \mathbf{g}_1 \quad [27.10]$$

As shown in Figure 27.6A (which is from fcc Ni on fcc NiO), we have assigned $\mathbf{g}_2$ to the smaller "lattice" spacing and tm indicates "translational moiré" fringes. The vector $\mathbf{g}_{tm}$ corresponds to a set of fringes with spacing d_{tm}, as shown by the following simple manipulation

$$d_{tm} = \frac{1}{g_{tm}} = \frac{1}{g_2 - g_1} = \frac{\frac{1}{g_2} \cdot \frac{1}{g_1}}{\frac{1}{g_1} - \frac{1}{g_2}}$$
$$= \frac{d_2 d_1}{d_1 - d_2} = \frac{d_1}{1 - \frac{d_2}{d_1}} \quad [27.11]$$

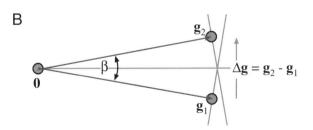

Figure 27.6. (A) Relationship between **g**-vectors for translational moiré fringes; (B) relationship for rotational moiré fringes.

27.5.B. Rotational Moiré Fringes

We follow the same procedure as above, but now the two **g**-vectors are identical in length and rotated through an angle β so that the new **g**-vector, $\mathbf{g}_{rm}$, has length $2g \sin \beta/2$, as shown in Figure 27.6B. The fringe spacing is then

$$d_{rm} = \frac{1}{g_{rm}} = \frac{1}{2g \sin \beta/2} = \frac{d}{2 \sin \beta/2} \quad [27.12]$$

27.5.C. General Moiré Fringes

If we use the same approach to locate $\mathbf{g}_{gm}$ (gm: "general moiré") we can readily show that, for small misorientation, the spacing d_{gm} of our fringes is given by

$$d_{gm} = \frac{d_1 d_2}{\left((d_1 - d_2)^2 + d_1 d_2 \beta^2 \right)^{\frac{1}{2}}} \quad [27.13]$$

27.6. EXPERIMENTAL OBSERVATIONS OF MOIRÉ FRINGES

Moiré fringes in TEM images were first reported early in the history of the microscope. They were used by Minter (1956) to identify a dislocation before lattice imaging was possible. Later, they were regarded as an imaging artifact which obscured the true dislocation structure in twist boundaries. Most recently there has been renewed interest due to the widespread development of thin films grown on different substrates.

You must be wary of the limitation or pitfall of using moiré fringes to learn about interfaces and defects. Moiré patterns result purely from the interference of two "sets of planes." Their appearance will be essentially the same even if the two "crystals" are not in contact.

In TEM the moiré patterns correspond to interference between a pair of beams, $\mathbf{g}_1$ and $\mathbf{g}_2$. If $\mathbf{g}_1$ is generated in the upper crystal and $\mathbf{g}_2$ in the lower, then each reflection $\mathbf{g}_1$ in crystal 1 acts as an incident beam for the lower crystal and produces a "crystal-2 pattern" around each $\mathbf{g}_1$ reflection, as shown in Figure 27.7A. This process is another example of double diffraction as discussed in Section 18.7. Figure 27.7A is from a pair of perfectly aligned but misfitting cubic crystals viewed along their common [001] zone axis; the pattern is indexed in Figure 27.7B. When we have many planes diffracting at a zone axis, as in this pattern, we expect to see crossed moiré fringes.

In the following three sections we will discuss examples of the use of moiré fringes.

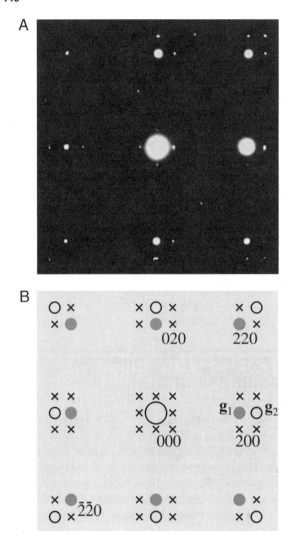

Figure 27.7. (A) Experimental diffraction pattern from perfectly aligned Ni and NiO. Brighter spots are from NiO, which has the larger lattice parameter. (B) Schematic which explains translational moiré fringes. Closed circles (g_1) correspond to crystal 1, open circles (g_2) to crystal 2 and × to double diffraction of g_1 beams by crystal 2. Only × reflections close to g_1 and g_2 have appreciable intensity.

27.6.A. Translational Moiré Patterns

When a continuous film is grown on a thick substrate, one question which is asked is: do the lattice parameters of the thin film correspond to the values of the same material in bulk form? For example, a thin film of a cubic material on a (001) substrate may be tetragonally strained so that the a_{film} lattice parameter is smaller than a_{bulk}, but the c_{film} parameter is larger. If the bulk material has its bulk lattice parameter, then the measurement of d_{tm}, the translational moiré spacing, can give a very accurate value for a_{film}. Furthermore, we can tilt the specimen 45° or 60° and deduce a value for c_{film} to estimate the tetragonal distortion directly.

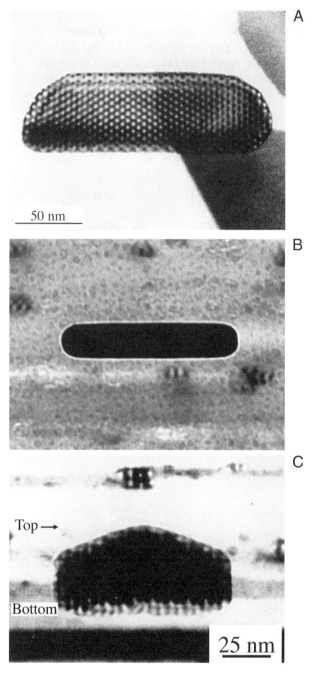

Figure 27.8. (A) The appearance of moiré fringes depends on the thickness of the specimen, as you can see where the edges of this island are inclined relative to the surface of the substrate. (B) The particle is too thick to show moiré fringes when edge on. (C) When this thick particle is tilted over, moiré fringes are seen at both the top and bottom.

Tilting the specimen can also give us information about misfitting islands, as illustrated in Figure 27.8. In this case, we see a hexagonal array of fringes when the two pseudo-hexagonal materials are viewed parallel to their common c-axis. The variation in the contrast of the moiré fringes around the edge of the particle occurs because the particle facets on inclined planes, as is confirmed when we tilt the specimen. In this system, when the islands are grown on different substrates, they may still grow as platelets. In Figures 27.8B and C the platelet is thick in the direction of the beam but, when tilted over, we again see moiré fringes. In particular, we can see moiré fringes at the *top* of the platelet.

> We know that this region is not in contact with the substrate, which reminds us that these fringes do not tell us about the interface structure!

27.6.B. Rotational Moiré Fringes

We often see rotational moiré fringes at twist boundaries, as illustrated for Si in Figure 27.9. A complicating factor is that the misfit may be accommodated by an array of dislocations having a periodicity which is related to the moiré-fringe spacing. The periodic strain field from the dislocations is, of course, only present if the two materials are in intimate contact.

27.6.C. Dislocations and Moiré Fringes

Since the moiré pattern can often be thought of as a magnified view of the "structure" of the materials, such patterns can be used to locate and give information on dislocations which are present in one material but not the other. We can form an image which contains information about the dislocation if it is associated with a terminating lattice plane in one material, but we don't actually "see" the dislocation.

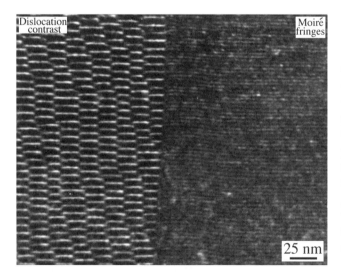

Figure 27.9. WBDF image of moiré fringes at a grain boundary showing very different contrast than the region containing dislocations.

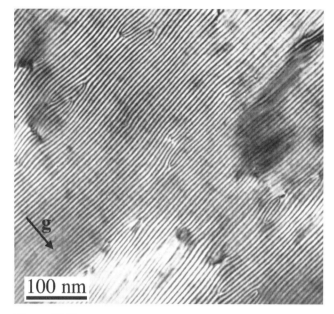

Figure 27.10. Moiré fringes reveal the presence of dislocations in a thin film of CoCa grown on a GaAs substrate. The (001) interface lies parallel to the specimen surface. Although the images contain much detail, most of it cannot readily be related to the structure of the defects.

This effect is illustrated in Figure 27.10; the image appears as a magnified view of the projection of the dislocation. This result can be deceptive, as you can see in Figure 27.11, where we have rotated the perfect grain slightly and changed its spacing.

> The images can always be related directly to the projected Burgers vector of the dislocation, but you must know which planes give rise to the fringes. So make some models and experiment.

This analysis even works if you have two or more terminating fringes, but don't put too much emphasis on the actual location of the fringes. Remember, the dislocation may not be parallel to the beam. Moiré fringes may be related to a dislocation in the plane of the interface, since these locally relax the misfit. One example of such an application comes from the work of Vincent (1969), who showed that as Sn islands grew on a thin film of SnTe, the moiré-fringe spacing around the perimeter of the islands gradually increased. Suddenly, the strain at the interface was so large that a dislocation was nucleated to relax the strain and the process began again. The analysis of the changes in moiré-fringe spacing is shown in Figure 27.12.

Since the spacing of moiré fringes essentially gives a magnified view of the misfit between aligned particles and a substrate, we can use them to measure the strain in

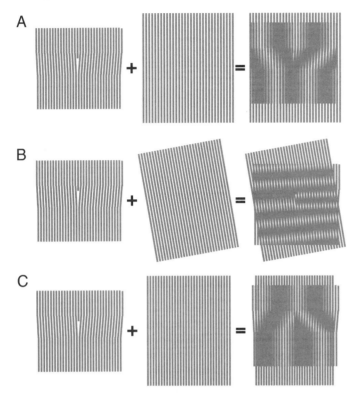

Figure 27.11. Schematic diagrams showing why moiré patterns from regions containing dislocations cannot be readily interpreted: (A) a dislocation image formed by interference between a regular lattice and one containing an extra half plane. (B) In comparison with (A), small rotation of the lattice of either grain can cause a large rotation of the dislocation fringes. (C) A small spacing change of either lattice can cause the dislocation image to reverse.

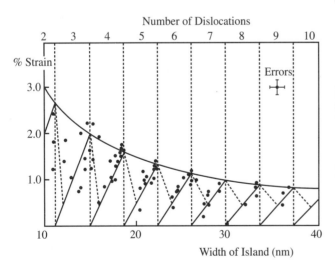

Figure 27.12. Moiré-fringe spacings can be used to monitor the change in lattice parameter as small islands of Sn grow in size on a thin film of SnTe. This plot shows how the strain (measured from the moiré-fringe spacing) can be related to the width of the misfitting island and then to the number of dislocations in the interface.

such particles. In its simplest form, in one dimension, the strain is given by

$$\varepsilon = \frac{a_1 - a_0}{a_0} \quad [27.14]$$

where a_1 and a_0 are the lattice spacings of the particle and substrate, respectively. You may need to modify this equation if the alignment is not simple cube-on-cube.

27.6.D. Complex Moiré Fringes

Since moiré fringes can occur whenever $\Delta\mathbf{g}$ is small enough to be included in the objective aperture, we can have a situation where the relative rotation is rather large (45° or even 90°) so that $\mathbf{g}_1$ and $\mathbf{g}_2$ correspond to different sets of planes. This is illustrated in Figure 27.13 for YBCO grains rotated 45° on a MgO substrate (Norton and Carter 1995). You can see that, as a bonus, the moiré fringes allow you to locate the 45° boundaries directly. Small rotations of the diffracting planes cause small rotations of $\mathbf{g}$ but large rotations of $\Delta\mathbf{g}$.

Two overlapping lattices produce a pattern of interference fringes which is much coarser than the original pattern and is very sensitive to differences in lattice spacing

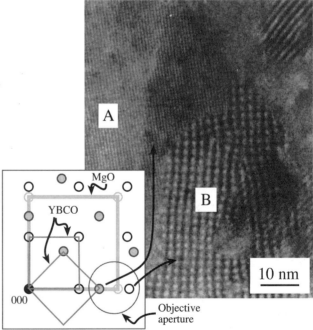

Figure 27.13. Moiré fringes formed when grains of YBCO grown on a single crystal of MgO are aligned to the substrate (B) or rotated through 45°(A); the spacing of the fringes is different so the position of the grain boundary can be identified. The circle in the DP shows the spots that cause the fringes. Small rotations of the fringes away from perfect alignment are exaggerated because the spots are close together.

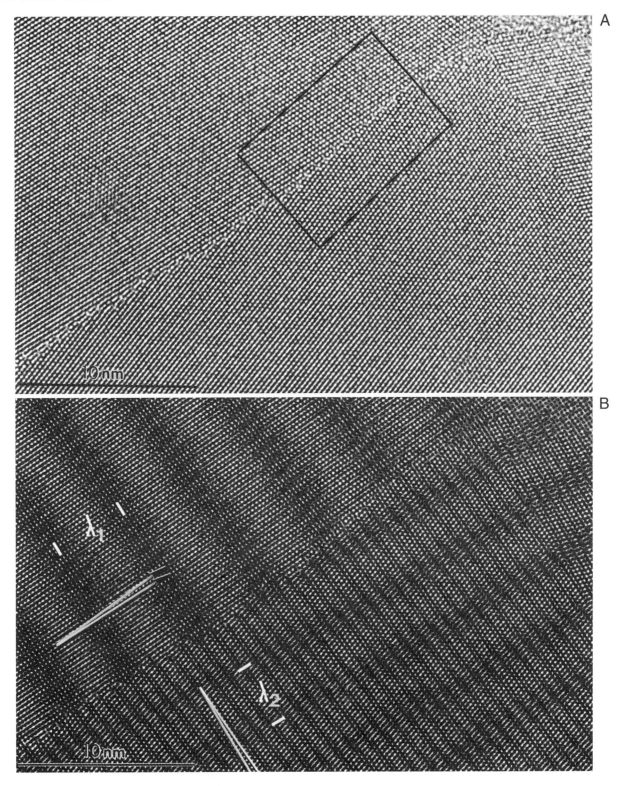

Figure 27.14. The use of "artificial" moiré fringes to analyze a special grain boundary in Al. (A) An experimental image. (B) The same image overlaid with a perfect-crystal lattice transparency, producing moiré fringes of different spacings, λ_1 and λ_2.

and relative orientations. We can use this sensitivity to provide an optical method for examining small rotations or lattice-parameter differences in HRTEM images. Make transparencies of the "distorted" image such as that shown in Figure 27.14A and a reference lattice; the reference lattice could be the perfect-crystal image or a template you have created on the computer. Now overlay the two and rotate/translate them relative to one another. You will have created a new artificial moiré image similar to that shown in Figure 27.14B, which was formed for a special grain boundary in Al by Hetherington and Dahmen (1992). This boundary is special, because one set of {111} planes in the upper grain is nearly normal to one set in the lower grain. How near is near? Hetherington and Dahmen overlapped their experimental image with a template which was drawn to have two sets of lines normal to one another. Overlaying the two images gave moiré fringes which were not quite perpendicular to one another. Careful measurements of the rotation and fringe spacing showed that the fringes in the experimental image were actually 89.3° apart, not 90°.

27.7. FRESNEL CONTRAST

We saw in Chapter 9 that we can use Fresnel-contrast images of holes in carbon films to correct the astigmatism of the objective lens. We'll now discuss how we can use this same contrast mechanism to learn more about particular features in the specimen. In the classical demonstration of Fresnel contrast using visible light, bright fringes can appear in the geometric shadow of an opaque mask, or dark fringes can appear in the illuminated region (e.g., Heavens and Ditchburn 1991). The complication introduced in the TEM version is that the "mask" is not opaque but simply has a different inner potential. Therefore, in any situation where the inner potential changes abruptly, we can produce Fresnel fringes if we image that region out of focus. Since we still focus on a plane which is close to the specimen, we are in the near-field or Fresnel regime. Now we extend this concept and say that

> Whenever we observe contrast only because we are forming an out-of-focus image, we are forming a Fresnel image.

Since we often study lines, planes, or platelets by this technique, we'll often see Fresnel fringes.

27.7.A. The Fresnel Biprism

We can demonstrate a particularly simple interference phenomenon by placing a wire at a position F on the optic axis, as shown in Figure 27.15A. Since the beam is narrow, the wire should be less than 1 μm in diameter and can be made of a drawn glass fiber coated with Cr or Au. If we apply ~10 V to the wire, it will bend the electron beam on either side in opposite directions. The resulting interference fringes can be recorded on a photographic film, as shown in Figure 27.15B. The wire here is acting as a beam splitter; we'll encounter it again when we discuss holography in Chapter 31. The visible-light analog is the prism. Notice how the wire acts to produce two virtual sources s_1 and s_2, which are D_s apart. Horiuchi (1994) gives the following equation to define a measure of the degree of spatial coherence, γ, which, as we discussed in Section 5.2 and Figure 5.13, is a function of the source size. Horuchi shows that

$$\gamma = \frac{I_{Max} - I_{Min}}{I_{Max} + I_{Min}} \qquad [27.15]$$

where I_{Max} is the intensity of the central fringe and I_{Min} is the intensity of the first minimum in Figure 27.15B.

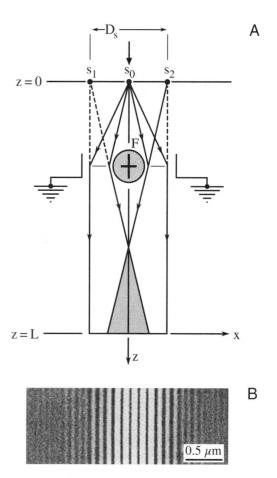

Figure 27.15. (A) A Fresnel biprism formed using a charged wire placed in the path of the beam; (B) the resulting interference fringes in the image.

27.7.B. Magnetic-Domain Walls

Although we'll discuss imaging magnetic materials in Chapter 31, it is appropriate here to consider briefly the similarity of Lorentz microscopy of magnetic-domain walls to other interference images. We know from our discussion of magnetic fields in the electron lens, in Chapter 6, that the Lorentz force on an electron with velocity **v** is proportional to **v**×**B**. If the sign of **B** is opposite in two adjacent domains, then the electrons will be deflected in opposite directions, as shown in Figure 27.16. The "converging" domain wall is remarkably similar to the electron interferometer. We can indeed produce a series of interference fringes. You should consult the original analysis of Boersch et al. (1960), but the basics are given by Hirsch et al. (1977) who show that the fringe spacing Δx is given by

$$\Delta x = \frac{\lambda(L + \ell)}{2\ell\beta_m} \qquad [27.16]$$

where β_m is the angle of deflection of the beam, λ is the electron wavelength, ℓ is the "source"-to-specimen distance, and L is the specimen-to-"detector" distance. Quotation marks are used to emphasize that these are "effective" distances like the "camera length." The value of Δx can be ~20 nm. You'll only see such interference fringes if you form the image using parallel illumination. We'll return to magnetic imaging in Chapter 31.

27.8. FRESNEL CONTRAST FROM VOIDS OR GAS BUBBLES

You might think that it would be difficult to image voids or small gas-filled cavities when there is no associated strain field, because voids or cavities do not scatter electrons. However, we can image holes which are fully enclosed inside the specimen by defocusing the image and observing the special form of phase contrast termed Fresnel contrast, which we introduced back in Sections 2.9 and 9.5. In principle, we can apply this technique to holes which contain a liquid or even a solid (i.e., a second phase). In the latter case, however, the Fresnel contrast is likely to be hidden by strain contrast in the specimen.

You can image small voids or gas bubbles in two ways:

- By orienting the region of interest so that **s** = 0; the cavity then reduces the "thickness" of material locally.
- By using Fresnel contrast, as explained by Wilkens (1975).

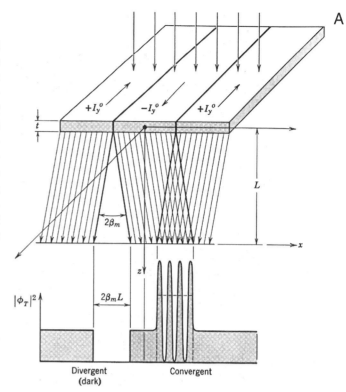

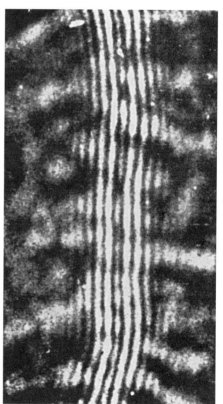

Figure 27.16. (A) Deflection of the electron beam by magnetic-domain walls; compare with Figure 27.15A. (B) Interference fringes from one such wall; compare with Figure 27.15B.

In the Fresnel technique, the image shows contrast whenever the objective lens is not focused on the bottom surface of the specimen.

| Fresnel-contrast images are always out of focus. |

Wilkens expresses the wave function as

$$\psi(t, \mathbf{r}') = \psi_0(t)[1 + \Delta_r(\mathbf{r}') + i\Delta_i(\mathbf{r}')] \quad [27.17]$$

Here, $\psi_0(t)$ is the wave function in the absence of the cavity; Δ_r and Δ_i are real functions which depend on:

- The location and dimensions of the cavity.
- The extinction distance and absorption parameter of the matrix (ξ_g and ξ_g').
- The potential difference, ΔV, between the inner potential of the matrix, V_0, and that of the cavity, V_c (it could be filled or empty).

In the case of thick foils where z_c, the size of the cavity in the direction of the beam, is $< 0.1\xi_g$, the wave function can be expressed as

$$\psi(t, \mathbf{r}') = \psi(t)[1 + i\Delta_i(\mathbf{r}')] \quad [27.18]$$

where Δ_i (using $w = s\xi_g$) is given by

$$\Delta_i = -\left(2\varepsilon_0 - \frac{1}{\xi_g}\frac{1}{(1+w)^{1/2}}\right)z_c(\mathbf{r}')p_i(z) \quad [27.19]$$

The difference in inner potential is included in ε_0, which is defined by the equation

$$\varepsilon_0 = -\frac{\Delta V}{E}k \quad [27.20]$$

Here, k is the magnitude of the wave vector and E is the energy of the electron beam. When the thickness dependence is damped out (the foil is thick), the intensity can be expressed quite simply as

$$|\psi(t, \mathbf{r}')|^2 = |\psi_0(t)|^2(1 + \Delta_i^2) \quad [27.21]$$

We can summarize some results from this study:

- When the image is in focus, the cavity is invisible so you have to view it out of focus to observe the Fresnel-fringe contrast.
- The contrast depends on the difference in the inner potential of the matrix and the cavity; we usually see the most contrast if the content of the cavity is vacuum, because then ε_0 is greatest.

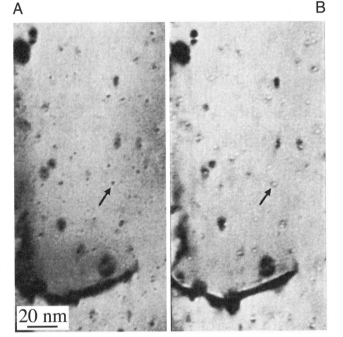

Figure 27.17. Fresnel contrast from He bubbles in Au. (A) Overfocus image. (B) Underfocus image.

- The contrast does depend on the wavelength of the electrons through both k and E.
- Cavities as small as 1–2 nm in diameter can be imaged using Δf values of 0.5 to 1.0 μm.
- In the case where $w = 0$ and $2\varepsilon_0 > \xi_g^{-1}$ (so Δ_i is < 0), if $\Delta f < 0$, the image is a bright dot surrounded by a dark fringe; if $\Delta f > 0$, the dot is dark and the fringe is bright.
- This is the same behavior as we saw in Figure 9.20, where we had a dark fringe at underfocus and a bright fringe at overfocus.

The contrast is illustrated in Figure 27.17. You should note that it is not the same as the black–white contrast from small precipitates discussed in Chapter 25. You'll find a more detailed analysis in the article by Rühle and Wilkens (1975).

27.9. FRESNEL CONTRAST FROM LATTICE DEFECTS

This topic is one which is receiving more attention as the computer and simulation programs become, respectively, more powerful and more user-friendly. The reason for this increased attention is clear, as Bursill *et al.* (1978) showed in their pioneering studying of Fresnel fringes from edge-on defects. They demonstrated that, if you take great care in determining all the electron optical parameters (particularly what defocus steps you are using), you can obtain new information on edge-on defects. The defect images are very sensitive to the model used to simulate them, but you will need other information for a full analysis. We'll refer you to their paper on the {100} platelets in diamond and concentrate on two more widely applicable situations, namely, end-on dislocations and edge-on grain boundaries; in both cases we now have many techniques, such as XEDS, EELS, and HRTEM, to complement the Fresnel-fringe studies.

27.9.A. Grain Boundaries

We might expect almost any grain boundary to show a localized change in the inner potential. However, following the original suggestion by Clarke (1979), Fresnel-contrast imaging has been used most extensively to study those interfaces which are thought to contain a thin layer of glass (Clarke 1979, Ness et al. 1986, Rasmussen et al. 1989, Rasmussen and Carter 1990). Part of the reason for this emphasis is simply that other techniques tend to give ambiguous results for such interfaces (Simpson et al. 1986).

When you use the Fresnel-fringe technique to study grain boundaries or analyze intergranular films, you must orient the boundary in the edge-on position so that you can probe the potential at the boundary. Later, in Section 29.11, we will consider the actual shape of this "potential well."

In a real TEM specimen, the grain boundary is likely to change in thickness even if only by a nanometer or so. Since the specimen will be quite thin, this change can give an appreciable contribution to the difference in the "effective inner potential" seen by the electron beam. You can defocus the image to see the Fresnel contrast shown in Figure 27.18.

The Fresnel-contrast technique can equally well be applied to phase boundaries, with perhaps the most thoroughly studied example being the Si/SiO_2 interface (Taftø et al. 1986, Ross and Stobbs 1991a). Since the details of the contrast are sensitive to the abruptness of the change in the inner potential, the technique can also produce information on this aspect of the interface (Ross and Stobbs 1991b). Nevertheless, you must always look for associated changes in the real geometry which can even occur when you're just forming Fresnel fringes from the edge of the specimen (Fukushima et al. 1974).

27.9.B. End-On Dislocations

We've just seen that we can detect Fresnel-fringe contrast from edge-on high-angle grain boundaries. We might then ask: is it possible to detect similar contrast from low-angle grain boundaries, i.e., grain boundaries which consist of arrays of distinct dislocations? It is indeed possible, as shown in the series of images from a tilt boundary in NiO

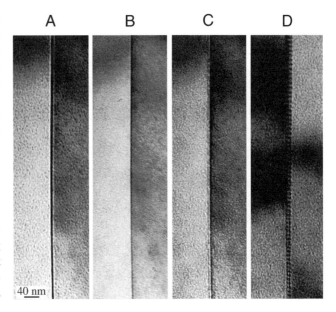

Figure 27.18. (A–D) A through-focus series of images from an edge-on GB showing the changes in Fresnel contrast. The image in (D) shows the boundary tilted over to reveal its periodic structure more clearly.

in Figure 27.19. Rühle and Sass (1984) analyzed this through-focus series, and images of other grain boundaries, by assuming that there is a change in $\Delta V(\mathbf{r})$ in the mean inner potential at the core of the dislocation. They proposed two models for $\Delta V(\mathbf{r})$. In model 1, when $r < r_0$

$$\Delta V(\mathbf{r}) = \Delta V_0 \left\{ 1 - e^{-\frac{(r-r_0)}{a}} \right\} \qquad [27.22]$$

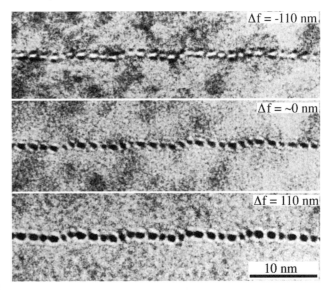

Figure 27.19. Series of experimental images recorded at different values of Δf for a low-angle grain boundary in NiO. Each white or black spot corresponds to one end-on dislocation.

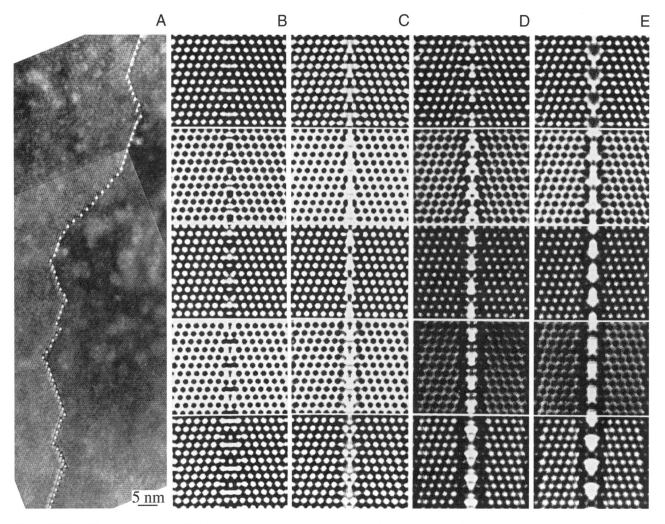

Figure 27.20. The structure of the (112) lateral twin boundary in a thin-foil of spinel consists of triangular prisms with a density lower than the bulk crystal. (A) Fresnel contrast is seen when these prism-like defects are imaged out of focus. (B–E) Simulated images; each column is a different model: (B) no ions removed, (C) ion sites in the prisms half occupied, (D,E) all ions removed. The defocus in each row increases from the top (–10 nm, –70 nm, –130 nm, –160 nm, –210 nm), and the thickness is 5.7 nm.

but when $r > r_0$

$$\Delta V(\mathbf{r}) = 0 \quad [27.23]$$

The constant a is $\sim 0.1 r_0$. In model 2

$$\Delta V(\mathbf{r}) = \Delta V_0 \exp\left(-r^2/r_0^2\right) \quad [27.24]$$

In both cases, ΔV_0 is negative. As an example of the quantities involved in these equations, if the Burgers vector of the dislocations is [110], Rühle and Sass found that $\Delta V_0 = 0.09 V_0$ for $r_0 = 3.2$ Å. They could not distinguish between the two models for $\Delta V(\mathbf{r})$, but two clear points come out of this study:

- You must know the inclination of your foil surface. If the lower surface is inclined to the horizontal, then thicker parts of the specimen can be much closer to the objective lens than in the thin area; you can do a quick calculation to prove this point.
- The inner potential at a dislocation core is different than the bulk value. You should expect the value of ΔV_0 to be influenced by a change in stoichiometry or impurity segregation.

Before leaving this topic we should point out that the inner potential at the grain boundary may not be uniform, perhaps because the width of the interface varies or the interface facets on a mesoscopic scale. Even then, you can still see Fresnel effects which relate to the periodicity in the grain boundary even if this periodicity is not associated with dislocations. A particularly clear example of such a variation is shown in Figure 27.20, where a twin boundary in spinel is essentially constructed of parallel triangular tubes; the inner potential inside the tube is much lower than the matrix value and the tubes are only about 1.2 nm high (Carter et al. 1987).

CHAPTER SUMMARY

Phase contrast will occur whenever we have more than one beam contributing to the image. The clue is: if you see fringes of any sort, then you are almost certainly observing a phase-contrast image. This conclusion even applies to stacking-fault fringes (Chapter 24) and thickness fringes (Chapter 23) in what are traditionally called two-beam diffraction-contrast images.

Phase-contrast images are widely used in three forms:

- Images which relate directly to the structural periodicity of the crystalline specimen.
- Moiré-fringe images.
- Fresnel-contrast images.

It is even possible for an image to show all three effects at the same time. So you must remember that phase-contrast effects don't just occur when you are forming high-resolution images. You will create Fresnel contrast whenever your specimen is thick or you are working out of focus. You should note that it is difficult, but not impossible, to be quantitative in your analysis of Fresnel fringes.

The usefulness of moiré fringes continues to surprise even experienced users of the TEM. However, you still have to exercise caution when interpreting what they are telling you about defects in your material.

The appearance of the Fresnel image varies with small changes in the thickness, orientation, or scattering factor of your specimen, and variations in the focus or astigmatism of the objective lens.

REFERENCES

Specific References

Boersch, H., Hamisch, H., Wohlleben, D., and Grohmann, K. (1960) *Z. Phys.* **167**, 72.
Bursill, L.A., Barry, J.C., and Hudson, P.R.W. (1978) *Phil. Mag.* **A37**, 789.
Carter, C.B., Elgat, Z., and Shaw, T.M. (1987) *Phil. Mag.* **55**, 1.
Clarke, D.R. (1979) *Ultramicroscopy* **4**, 33.
Fukushima, K., Kawakatzu, H., and Fukami, A. (1974) *J. Phys.* **D7**, 257.
Heavens, O.S. and Ditchburn, R.W. (1991) *Insight into Optics,* p. 73, John Wiley & Sons, New York.
Hetherington, C.J.D. and Dahmen, U. (1992) *Scanning Microscopy Supplement* **6**, p. 405, Scanning Microscopy International, AMF O'Hare, Illinois.
Hirsch, P.B., Howie, A., Nicholson, R.B., Pashley, D.W., and Whelan, M.J. (1977) *Electron Microscopy of Thin Crystals,* 2nd edition, Krieger, Huntington, New York.
Horiuchi, S. (1994) *Fundamentals of High-Resolution Transmission Electron Microscopy,* North-Holland, Amsterdam.

Krivarnek, O.L. and Rez, P. (1980) Proc. 38th Ann. EMSA Meeting (Ed. G.W. Bailey), p. 170, Claitors, Baton Rouge, Louisiana.
Minter, J.W. (1956) *Proc. Roy. Soc.* (London) **A236**, 119.
Ness, J.N., Stobbs, W.M., and Page, T.F. (1986) *Phil. Mag.* **54**, 679.
Norton, M.G. and Carter, C.B. (1995) *J. Mater. Sci.* **30**, 381.
Rasmussen, D.R. and Carter, C.B. (1990) *Ultramicroscopy* **32**, 337.
Rasmussen, D.R., Simpson, Y.K., Kilaas, R., and Carter, C.B. (1989) *Ultramicroscopy* **30**, 52.
Ross, F.M. and Stobbs, W.M. (1991a) *Phil. Mag.* **A63**, 1.
Ross, F.M. and Stobbs, W.M. (1991b) *Phil. Mag.* **A63**, 37.
Rühle, M. and Sass, S.L. (1984) *Phil. Mag.* **A49**, 759.
Rühle, M. and Wilkens, M. (1975) *Crystal Lattice Defects* **6**, 129.
Simpson, Y.K., Carter, C.B., Morrissey, K.J., Angelini, P., and Bentley, J. (1986) *J. Mater. Sci.* **21**, 2689.
Taftø, J., Jones, R.H., and Heald, S.M. (1986) *J. Appl. Phys.* **60**, 4316.
Vincent, R. (1969) *Phil. Mag.* **19**, 1127.
Wilkens, M. (1975) in *Electron Microscopy in Materials Science,* **II** (Eds. U. Valdrè and E. Ruedl), p. 647, CEC, Brussels.

High-Resolution TEM

28

28.1. The Role of an Optical System	459
28.2. The Radio Analogy	459
28.3. The Specimen	461
28.4. Applying the WPOA to the TEM	462
28.5. The Transfer Function	462
28.6. More on $\chi(u)$, sin $\chi(u)$, and cos $\chi(u)$	463
28.7. Scherzer Defocus	465
28.8. Envelope Damping Functions	466
28.9. Imaging Using Passbands	467
28.10. Experimental Considerations	468
28.11. The Future for HRTEM	469
28.12. The TEM as a Linear System	470
28.13. FEGTEMs and the Information Limit	470
28.14. Some Difficulties in Using an FEG	473
28.15. Selectively Imaging Sublattices	475
28.16. Interfaces and Surfaces	476
28.17. Incommensurate Structures	478
28.18. Quasicrystals	479
28.19. Single Atoms	480

CHAPTER PREVIEW

We will now rethink what we mean by a TEM, in a way that is more suitable for HRTEM, where the purpose is to maximize the useful detail in the image. (Note the word *useful* here.) You should think of the microscope as an optical device which transfers information from the specimen to the image. The optics consist of a series of lenses and apertures aligned along the optic (symmetry) axis. What we would like to do is to transfer *all* the information from the specimen to the image, a process known as mapping. There are two problems to overcome and we can never be completely successful in transferring *all* the information. As you know from Chapter 6, the lens system is not perfect so the image is distorted and you lose some data (Abbe's theory). The second problem is that we have to interpret the image using an atomistic model for the material. Ideally, this model will include a full description of the atomic potential and the bonding of the atoms, but we don't know that either. We will also need to know exactly how many atoms the electron encountered on its way through the specimen. So most of our task will be concerned with finding the best compromise and producing models for the real situation. To conclude our discussion of the theory, we will introduce the language of *information theory,* which is increasingly used in HRTEM. We close the chapter with a review of the experimental applications of HRTEM to include periodic and nonperiodic materials, mixtures of the two, or just single atoms.

High-Resolution TEM

28

28.1. THE ROLE OF AN OPTICAL SYSTEM

What the microscope does is to transform each point on the specimen into an extended region in the final image. Since each point on the specimen may be different, we describe the specimen by a specimen function, $f(x,y)$. The extended region in the image which corresponds to the point (x,y) in the specimen is then described as $g(x,y)$, as shown schematically in Figure 28.1; note that both f and g are functions of x and y.

If we consider two nearby points, A and B, they will produce two overlapping images, g_A and g_B. If we extend this argument, we can see that each point in the image has contributions from many points in the specimen. We express this result mathematically by

$$g(\mathbf{r}) = \int f(\mathbf{r}') h(\mathbf{r} - \mathbf{r}') d\mathbf{r}' \qquad [28.1]$$

$$= f(\mathbf{r}) \otimes h(\mathbf{r} - \mathbf{r}') \qquad [28.2]$$

Here, $h(\mathbf{r} - \mathbf{r}')$ is a weighting term telling us how much each point in the specimen contributes to each point in the image.

> Since $h(\mathbf{r})$ describes how a point spreads into a disk, it is known as the point-spread function or smearing function, and $g(\mathbf{r})$ is called the convolution of $f(\mathbf{r})$ with $h(\mathbf{r})$.

Spence (1988) calls $h(\mathbf{r})$ the impulse response function, and notes that it can only apply to small patches of specimen which lie in the same plane and are close to the optic axis. The symbol $\otimes$ indicates that the two functions, f and h, are "folded together" (multiplied and integrated) or "convoluted with one another."

28.2. THE RADIO ANALOGY

We can compare this imaging process with the task of recording an orchestra on a record/tape/CD or even transmitting to the brain directly or via a radio. We want to hear the loud drum and quiet flute (large amplitude and small amplitude); we want to hear the high note on the violin and the low note on the double bass (high frequency and low frequency). Our audio amplifier has limits on both the low and high frequencies, so we won't achieve perfect reproduction. The importance of amplitude is obvious (more of this later), but how do we define frequency in a TEM image? High frequency in audio is related to $1/t$; frequencies in lattice images are related to $1/x$. So the high spatial frequencies simply correspond to small distances. What we are looking for in high-resolution work are the high spatial frequencies. Notice our use of high/low and large/small.

> High resolution requires high spatial frequencies.

Figure 28.2 shows two points A and B in the specimen and their disk images on the screen. We see disks (see our discussion of the Rayleigh disk in Chapter 6) because the lens system is not perfect. We can also write $g(x,y)$, the intensity of an image at point (x,y), as $g(\mathbf{r})$, and in the simplest case, these disks have uniform intensity. We can always represent any function in two dimensions as a sum of sine waves

$$g(x,y) = \sum_{u_x, u_y} G(u_x, u_y) \exp\left(2\pi i \left(xu_x + yu_y\right)\right) \qquad [28.3]$$

$$g(x,y) = \sum_{\mathbf{u}} G(\mathbf{u}) \exp\left(2\pi i \, \mathbf{u} \cdot \mathbf{r}\right) \qquad [28.4]$$

Here $\mathbf{u}$ is a reciprocal-lattice vector, the spatial frequency for a particular direction. We have expressed $g(\mathbf{r})$ in terms of a combination of the possible values of $G(\mathbf{u})$,

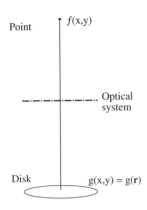

Figure 28.1. An optical system transforms a point in the specimen (described by $f(x,y)$) into a disk in the image described by $g(x,y)$. The intensity in the image at point (x,y) can be described by the function $g(x,y)$ or $g(\mathbf{r})$. It has a unique value for each value of (x,y) so we say that $g(\mathbf{r})$ is a representation of the image.

where $G(\mathbf{u})$ is known as the Fourier transform of $g(\mathbf{r})$. We can now define two other Fourier transforms:

$F(\mathbf{u})$ is the Fourier transform of $f(\mathbf{r})$,
and
$H(\mathbf{u})$ is the Fourier transform of $h(\mathbf{r})$.

Since $h(\mathbf{r})$ tells us how information in real space is transferred from the specimen to the image, $H(\mathbf{u})$ tells us how information (or contrast) in $\mathbf{u}$ space is transferred to the image.

$H(\mathbf{u})$ is the contrast transfer function.

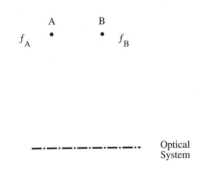

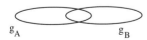

Figure 28.2. Two points, f_A and f_B, in the specimen produce two disks, g_A and g_B, in the image.

Now these three Fourier transforms are related by

$$G(\mathbf{u}) = H(\mathbf{u})\,F(\mathbf{u}) \qquad [28.5]$$

So a convolution in real space (equation 28.1) gives multiplication in reciprocal space (equation 28.5).

The factors contributing to $H(\mathbf{u})$ include:
Apertures $\rightarrow$ The aperture function $A(\mathbf{u})$
Attenuation of the wave $\rightarrow$ The envelope function $E(\mathbf{u})$
Aberration of the lens $\rightarrow$ The aberration function $B(\mathbf{u})$

We write $H(\mathbf{u})$ as the product of these three terms

$$H(\mathbf{u}) = A(\mathbf{u})\,E(\mathbf{u})\,B(\mathbf{u}) \qquad [28.6]$$

The aperture function says that the objective diaphragm cuts off all values of $\mathbf{u}$ (spatial frequencies) greater than (higher than) some selected value governed by the radius of the aperture. The envelope function has the same effect but is a property of the lens itself, and so may be either more or less restricting than $A(\mathbf{u})$. $B(\mathbf{u})$ is usually expressed as

$$B(\mathbf{u}) = \exp\left(i\chi(\mathbf{u})\right) \qquad [28.7]$$

The term $\chi(\mathbf{u})$ can be written as

$$\chi(\mathbf{u}) = \pi\,\Delta f\,\lambda\,u^2 + \tfrac{1}{2}\pi\,C_s\,\lambda^3\,u^4 \qquad [28.8]$$

We will give a crude derivation of this equation in Section 28.6. It builds on the concepts we discussed in Chapter 6 when we examined the origin of C_s.

> $\Delta f > 0$ is known as overfocus. It means we have focused the objective lens on a plane above the specimen. (By above, we mean before the electrons reach the specimen; the story is the same if the microscope is upside down!)

Summarizing so far: High spatial frequencies correspond to large distances from the optic axis in the DP. The rays which pass through the lens at these large distances are bent through a larger angle by the objective lens. They are not focused at the same point by the lens, because of spherical aberration, and thus cause a spreading of the point in the image. The result is that the objective lens magnifies the image but confuses the fine detail. The resolution we require in HRTEM is limited by this "confusion."

- Each point in the specimen plane is transformed into an extended region (or disk) in the final image.

28 ■ HIGH-RESOLUTION TEM

- Each point in the final image has contributions from many points in the specimen.

> What we need for our analysis to "work" is a "linear relationship between the image and the weak specimen potential."

We now have to go back and look at how we can represent the specimen. That is, what is $f(\mathbf{r})$ in equation 28.1? (We'll use the coordinates $\mathbf{r}$ and x,y interchangeably in this discussion; the former is more compact, but we can extend the latter notation to emphasize the possibility of a z component.)

28.3. THE SPECIMEN

Since we are using a TEM, we call the specimen function, $f(\mathbf{r})$, the specimen transmission function. Here you have to be very careful to remember that we are going to use a model to represent the specimen and the model will make certain assumptions. A general model would describe $f(\mathbf{r})$ as

$$f(x,y) = A(x,y)\exp(-i\phi_t(x,y)) \quad [28.9]$$

where $A(x,y)$ is the amplitude (not the aperture function) and $\phi_t(x,y)$ is the phase which depends on the thickness of the specimen.

For our application to HRTEM we simplify our model further by setting $A(x,y) = 1$; i.e., we set the incident-wave amplitude to be unity. We can show that the phase change only depends on the potential $V(x,y,z)$ which the electron sees as it passes through the specimen, by the following argument (Van Dyck 1992). To do this we assume that the specimen is so thin that we can write down a projected potential $V_t(x,y)$, with t being the thickness of the specimen, as usual

$$V_t(x,y) = \int_0^t V(x,y,z)dz \quad [28.10]$$

What we are doing is creating a two-dimensional projection of the crystal structure; this approach is critical to much of our interpretation of HRTEM images.

We can relate the wavelength, λ, of the electrons in vacuum to the energy. (Ideally, λ should have its relativistic value, but the principle is correct.)

$$\lambda = \frac{h}{\sqrt{2meE}} \quad [28.11]$$

(We'll give the analysis in a simple nonrelativistic form for simplicity.) When the electrons are in the crystal, λ is changed to λ'

$$\lambda' = \frac{h}{\sqrt{2me(E+V(x,y,z))}} \quad [28.12]$$

so we can say that, when passing through a slice of material of thickness dz, the electrons experience a phase change given by

$$d\phi = 2\pi\frac{dz}{\lambda'} - 2\pi\frac{dz}{\lambda} \quad [28.13]$$

$$d\phi = 2\pi\frac{dz}{\lambda}\left(\frac{\sqrt{E+V(x,y,z)}}{\sqrt{E}} - 1\right) \quad [28.14]$$

$$d\phi = 2\pi\frac{dz}{\lambda}\left(\left(1 + \frac{V(x,y,z)}{E}\right)^{\frac{1}{2}} - 1\right) \quad [28.15]$$

$$d\phi \simeq 2\pi\frac{dz}{\lambda}\frac{1}{2}\frac{V(x,y,z)}{E} \quad [28.16]$$

$$d\phi \simeq \frac{\pi}{\lambda E}V(x,y,z)dz \quad [28.17]$$

$$d\phi \simeq \sigma V(x,y,z)dz \quad [28.18]$$

So the total phase shift is indeed dependent only on $V(x,y,z)$ since

$$d\phi \simeq \sigma \int V(x,y,z)dz = \sigma V_t(x,y) \quad [28.19]$$

where $V_t(x,y)$ is the potential projected in the z-direction.

We call σ the interaction constant. It is not a scattering cross section, but is another expression for the elastic interaction we discussed in Chapter 3. It tends to a constant value as V increases, since the energy of the electron is proportional to E or λ^{-1} (i.e., changes in the two variables, λ and E, tend to compensate for one another).

Now, we can take account of absorption by including a function $\mu(x,y)$ (Cowley 1992) so that our specimen transfer function $f(x,y)$ is given by

$$f(x,y) = \exp[-i\sigma V_t(x,y) - \mu(x,y)] \quad [28.20]$$

The effect of this model is that, apart from $\mu(x,y)$, we have represented the specimen as a "phase object." This is known as the phase-object approximation, or POA. We are actually lucky because the absorption will usually be small in the regime where the rest of the approximation holds.

> In general, the phase-object approximation only holds for thin specimens.

We can simplify the model further if the specimen is *very* thin so that $V_t(x,y)$ is <<1. Then we expand the exponential function, neglecting μ and higher-order terms, so that $f(x,y)$ becomes

$$f(x,y) = 1 - i\sigma V_t(x,y) \quad [28.21]$$

Now we have reached the weak-phase-object approximation, or the WPOA. We see that the WPOA essentially says that, for a very thin specimen, the amplitude of a transmitted wave function will be linearly related to the projected potential of the specimen. Note that in this model the projected potential is taking account of variations in the z-direction, and is thus very different for an electron passing through the center of an atom compared to one passing through its outer regions.

Fortunately, there are many software packages that allow us to calculate what an image will look like for a particular specimen geometry. However, you must always remember that a model has been used to represent the specimen and have a clear understanding of its limits. To emphasize this last point, bear in mind that the WPOA fails for an electron wave passing through the center of a single uranium atom! As a second example, Fejes (1977) has shown that, for the complex oxide $Ti_2Nb_{10}O_{27}$, the WPOA is only valid if the specimen thickness is < 6 Å! The good news is that the approach appears to be more widely applicable than these particular estimates would suggest.

28.4. APPLYING THE WPOA TO THE TEM

So far, our treatment has been quite general, but now we will use our WPOA model. If we use the expression for $f(\mathbf{r})$ given by equation 28.21, then equation 28.2 tells us that the wave function as seen in the image is given by

$$\psi(x,y) = \left[1 - i\sigma V_t(x,y)\right] \otimes h(x,y) \quad [28.22]$$

If we represent $h(x,y)$ as cos (x,y) + i sin (x,y), then $\psi(x,y)$ becomes

$$\psi(x,y) = 1 + \sigma V_t(x,y) \otimes \sin(x,y) \\ - i\sigma V_t(x,y) \otimes \cos(x,y) \quad [28.23]$$

and the intensity is given by

$$I = \psi\psi^* = |\psi|^2 \quad [28.24]$$

Multiplying this out and neglecting terms in σ^2, because σ is small, we find that

$$I = 1 + 2\sigma V_t(x,y) \otimes \sin(x,y) \quad [28.25]$$

Knowing this result we can say that, in the WPOA, only the imaginary part of $B(\mathbf{u})$ in equation 28.7 contributes to the intensity in equation 28.24 (because it gives the imaginary part of $h(x,y)$). Therefore, we can set $B(\mathbf{u}) = 2 \sin \chi(\mathbf{u})$ rather than $\exp(i\chi(\mathbf{u}))$.

We can now define a new quantity, $T(\mathbf{u})$, which we'll call the transfer function to distinguish it from $H(\mathbf{u})$. It's given by

$$T(\mathbf{u}) = A(\mathbf{u}) E(\mathbf{u}) 2 \sin \chi(\mathbf{u}) \quad [28.26]$$

Note that $T(\mathbf{u})$ is not identical but is closely related to $H(\mathbf{u})$, which we defined in equation 28.6. The "2" in equation 28.26 is the "2" in equation 28.25 and arises because we are interested in the intensity in the beam, and therefore we multiplied ψ by its complex conjugate in equation 28.25. You may also see authors use a negative sign in equation 28.26 (in particular, in Reimer 1993). This has the effect of inverting the graph of $B(\mathbf{u})$ versus $\mathbf{u}$ and making $B(\mathbf{u})>0$ for positive phase contrast.

A note on terminology. You will often see $T(\mathbf{u})$ rather than $H(\mathbf{u})$ called the contrast transfer function in the HRTEM literature. The terminology comes from the analysis of the imaging process for incoherent light in visible-light optics. With incoherent illumination, $T(\mathbf{u})$ and $H(\mathbf{u})$ are identical. The smearing function (point-spread function) for that case is the Fourier transform of the contrast transfer function (the CTF). The equation describing $T(\mathbf{u})$ was derived for the situation where we have *coherent* imaging. For incoherent light the smearing function would be

$$\cos^2(x,y) + \sin^2(x,y) \quad [28.27]$$

which is just unity. So the CTF in HRTEM would be different from $T(\mathbf{u})$, and therefore we will call $T(\mathbf{u})$ the *objective lens transfer function*.

28.5. THE TRANSFER FUNCTION

You must note two things here. First, as we just said, the transfer function, $T(\mathbf{u})$, formulation applies to any specimen, and second, $T(\mathbf{u})$ is *not* the "contrast transfer function" of HRTEM. The problem with this formulation is that the image wave function is not an observable quantity! What we observe in an image is contrast, or the equivalent in optical density, current readout, etc., and this is not linearly related to the object wave function. Fortunately, there is a linear relation involving observable quantities under the special circumstances where the specimen acts as a "weak-phase object."

If the specimen acts as a weak-phase object, then the transfer function $T(\mathbf{u})$ is sometimes called the "contrast transfer function," because there is no amplitude contribution, and the output of the transmission system is an observable quantity (image contrast). The transfer function appropriate for this image formation process has the form which we derived above (equation 28.26), and if we ignore $E(\mathbf{u})$

$$T(\mathbf{u}) = 2A(\mathbf{u}) \sin \chi(\mathbf{u}) \quad [28.28]$$

where we know that $A(\mathbf{u})$ is the aperture function and might call $\chi(\mathbf{u})$ the phase-distortion function.

In other words, the phase-distortion function has the form of a phase shift expressed as $2\pi/\lambda$ times the path difference traveled by those waves affected by spherical aberration (C_s), defocus (Δz), and astigmatism (C_a).

Assuming that astigmatism can be properly corrected, the phase-distortion function is the sum of two terms. If the contrast transfer function is now compared to the phase-distortion function, a number of observations can be made. Note that the contrast transfer function is oscillatory; there are "bands" of good transmission separated by "gaps" (zeros) where no transmission occurs.

The contrast transfer function shows maxima (meaning maximum transfer of contrast) whenever the phase-distortion function assumes multiple odd values of $\pm\pi/2$. Zero contrast occurs for $\chi(\mathbf{u})$ = multiple of $\pm\pi$.

> When $T(\mathbf{u})$ is negative, positive phase contrast results, meaning that atoms would appear dark against a bright background. When $T(\mathbf{u})$ is positive, negative phase contrast results, meaning that atoms would appear bright against a dark background. When $T(\mathbf{u}) = 0$, there is no detail in the image for this value of $\mathbf{u}$.

The reason for this behavior is that the phase shift due to diffraction is $-\pi/2$. If a diffracted beam is further phase-shifted by $-\pi/2$, it subtracts amplitude from the forward-scattered beam, causing atoms to appear dark (positive contrast). If the same beam is instead phase-shifted by $+\pi/2$, it adds amplitude to the forward-scattered beam (they are "in phase"), causing atoms to appear bright (negative contrast).

28.6. MORE ON $\chi(\mathbf{u})$, SIN $\chi(\mathbf{u})$, AND COS $\chi(\mathbf{u})$

The ideal form of $T(\mathbf{u})$ would be a constant value as $\mathbf{u}$ increases, as shown in Figure 28.3; $T(\mathbf{u})$ must be zero at $\mathbf{u} = 0$ but, since small values of $\mathbf{u}$ correspond to very large values of x (i.e., long distances in the specimen), this is not a problem. If $T(\mathbf{u})$ is large, it means that information with a periodicity or spatial frequency corresponding to that value of $\mathbf{u}$ will be strongly transmitted, i.e., it will appear in the image. What we then need is that the different values of $\mathbf{u}$ give the same contrast. Then all the atoms in a crystal appear as black spots, say, rather than some as black spots and others as white spots; if the latter occurred, interpretation would be difficult!

> $T(\mathbf{u})$ becomes zero again at $\mathbf{u} = \mathbf{u}_1$; what we would like is for $\mathbf{u}_1$ to be as large as possible. If $T(\mathbf{u})$ crosses the $\mathbf{u}$-axis, the sign of the transfer function reverses. This means that $\mathbf{u}_1$ defines the limit at which our image may be quite directly interpreted; it is a very important parameter.

We will now go through a simple exercise to produce an expression for $\chi(\mathbf{u})$. If we combine the effects of the spherical aberration (equation 6.15) and the defocus (equation 11.19) of the objective lens, we find that a point at the specimen will actually be imaged as a disk with radius $\delta(\theta)$

$$\delta(\theta) = C_s \theta^3 + \Delta f \theta \quad [28.29]$$

The rays which pass through the objective lens at angle θ are not focused to the same point due to the spherical aberration of the objective lens and the finite value of Δf. If we only had one value of θ, we would still be all right! Of course, we have a range of values, so we average (integrate) these with respect to θ to give

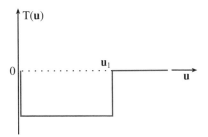

Figure 28.3. The ideal form of the transfer function, $T(\mathbf{u})$. In this example $T(\mathbf{u})$ is large and negative between $\mathbf{u} = 0$ and $\mathbf{u} = \mathbf{u}_1$.

$$D(\theta) = \int_0^\theta \delta(\theta)d\theta = \frac{C_s\theta^4}{4} + \Delta f\frac{\theta^2}{2} \quad [28.30]$$

Now, Bragg's Law tells us that

$$2d \sin \theta_B = n\lambda \quad [28.31]$$

or, since θ_B is small

$$2\theta_B \cong \lambda g \quad [28.32]$$

So, we can replace θ in equation 28.30 with λu where **u** is a general reciprocal lattice vector. Remember that the scattering angle is $2\theta_B$, not θ_B.

We are interested in the phase $\chi(\mathbf{u})$, so we write

$$\chi(\mathbf{u}) = \text{phase} = \frac{2\pi}{\lambda}D(\mathbf{u}) = \frac{2\pi}{\lambda}\left(C_s\frac{\lambda^4 u^4}{4} + \Delta f\frac{\lambda^2 u^2}{2}\right) \quad [28.33]$$

and we have

$$\chi = \pi \Delta f \lambda u^2 + \frac{1}{2}\pi C_s \lambda^3 u^4 \quad [28.34]$$

Clearly, $\sin \chi(\mathbf{u})$ will be a complicated curve which will depend on the values of C_s (the lens quality), λ (the accelerating voltage), Δf (the defocus value you choose to form the image), and **u** (the spatial frequency). (When you are ready for a more rigorous derivation of equation 28.34, see Spence, 1988, Section 3.3; most of us just start with equation 28.34 and this reasonable justification.)

The best way to appreciate the importance of χ is to use one of the simulation packages discussed in Chapter 29 and vary each of the parameters one by one. The plot of $T(\mathbf{u})$ (= $2 \sin \chi$) versus **u**, shown in Figure 28.4, illustrates the main features. The curve has been drawn for C_s = 1 mm, E_0 = 200 keV, and a defocus value of –58 nm.

The important features of this curve are shown in Figures 28.4–28.6:

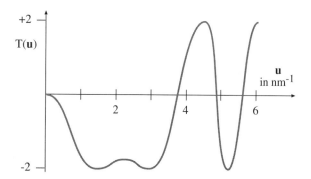

Figure 28.4. A plot of $T(\mathbf{u})$ versus **u** (C_s = 1 mm, E_0 = 200 keV, Δf = –58 nm).

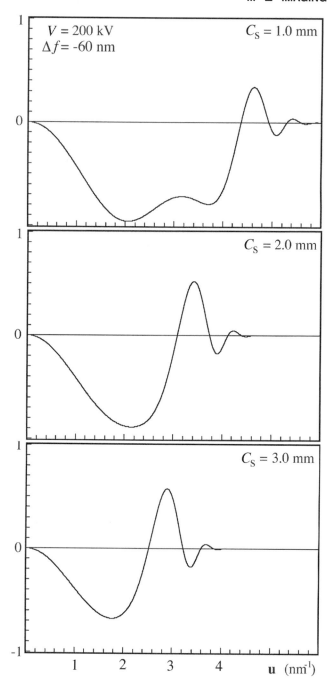

Figure 28.5. A series of $\sin \chi$ curves calculated for different values of C_s. Remember $2 \sin \chi = T(\mathbf{u})$.

- $\sin \chi$ starts at 0 and decreases. When **u** is small, the Δf term dominates.
- $\sin \chi$ first crosses the **u**-axis at $\mathbf{u}_1$ and then repeatedly crosses the **u**-axis as **u** increases.
- χ can continue forever but, in practice, it is modified by other functions, which we discuss in Section 28.8.

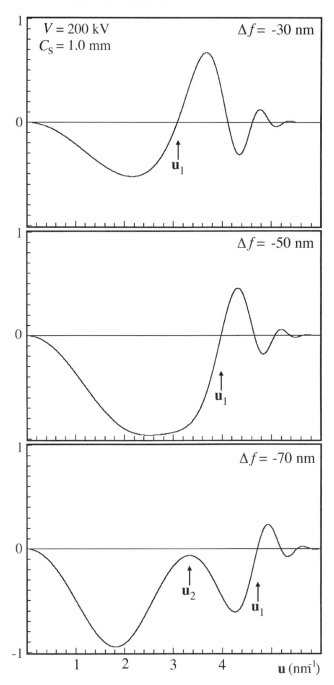

Figure 28.6. A series of sin χ curves calculated for different values of Δf.

Once you've selected your microscope and its objective lens, you have fixed C_s although C_s does depend to some extent on the λ you choose. The curve of $T(\mathbf{u})$ versus $\mathbf{u}$ does not depend on your specimen. Figure 28.5 shows a series of sin χ curves for an imaginary 200-kV microscope where C_s has been changed. In each case, the "best" curve (we'll discuss this in a moment) has been chosen. You can appreciate that the smaller C_s values give the larger $\mathbf{u}_1$ values; so a small C_s means we can achieve a higher spatial resolution.

If instead we fix C_s and again plot the best curves, but this time varying λ, you can see that the smallest value of λ allows us to achieve a higher spatial resolution. The result is not surprising; we want a small C_s and a small λ or a high voltage. So we choose the microscope to optimize C_s and λ. Now we only have Δf to vary. The set of curves shown in Figure 28.6 illustrates the effect of varying Δf. Notice that the bump in the curve at $\mathbf{u}_2$ will eventually increase as Δf increases until it crosses the $\mathbf{u}$-axis so that $\mathbf{u}_1$ is suddenly much smaller. If we just make Δf smaller, then $\mathbf{u}_1$ steadily decreases. In the next section we will discuss the optimum value for Δf.

> High spatial frequencies ⇒ large diffraction angles ⇒ larger effect of objective lens (C_s).
> So for a large objective aperture semiangle β, the $β^4$ term wins, i.e., C_s wins, but you can vary Δf.

28.7. SCHERZER DEFOCUS

The presence of zeros in the contrast transfer function means that we have gaps in the output spectrum which do not contribute to the output signal: it's as if these frequencies were filtered out. Obviously, the best transfer function is the one with the fewest zeros, which would be the case for a perfect lens, for example. What Scherzer did back in 1949 was to notice that the transfer function could be optimized by balancing the effect of spherical aberration against a particular negative value of Δf. This value has come to be known as "Scherzer defocus," $Δf_{Sch}$, which occurs at

$$Δf_{Sch} = -1.2 \left(C_s \lambda\right)^{\frac{1}{2}} \qquad [28.35]$$

At this defocus (which we'll derive below) all the beams will have nearly constant phase out to the "first crossover" of the zero axis. This crossover point is defined as the instrumental resolution limit. This is the best performance that can be expected from a microscope unless we use sophisticated image processing schemes to extract more information. In other words, this is not the information limit but it is the limit where we can use nearly intuitive arguments to interpret what we see. Again, as we discussed in Chapter 6 when we defined image resolution, you will see other authors give different values for the constant rather than the 1.2 given in equation 28.35; remember that this number is a calculated value, so it does depend on the details of your approximations.

This definition of resolution has new implications. The Rayleigh criterion which we used in Chapter 6 was only concerned with our ability to distinguish closely spaced object points. Our new definition requires a flat response in the object spectrum, and the goal is to have as many beams as possible being transferred through the optical system with identical phase, i.e., within the flat response regime. This is the underlying principle governing phase-contrast imaging in HRTEM.

> A TEM image with detail of 0.66 Å was demonstrated in 1970 when the *interpretable* resolution was about 3.3 Å. So just because you can see detail in the image does not mean that you can gain useful information about your specimen.

The closest we can get to the ideal curve in Figure 28.6 occurs when $\chi(\mathbf{u})$ is close to $-120°$; then $\sin \chi$ will be near -1 when χ is between $-120°$ and $-60°$. We know that when $\chi = \pi$, $\sin \chi = 0$, so we want $\sin \chi$ to be as large as possible over a large range of $\mathbf{u}$. $\sin \chi$ will be a nearly flat function if $d\chi/du$ is zero. So we look for the value of Δf when $d\chi/du$ is zero and χ is $-120°$ (you should consider why we choose this value of χ). Differentiating equation 28.29 gives

$$\frac{d\chi}{du} = 2\pi \Delta f \lambda u + 2\pi C_s \lambda^3 u^3 \quad [28.36]$$

Set the left-hand term equal to 0

$$0 = \Delta f + C_s \lambda^2 u^2 \quad [28.37]$$

When $\chi = -120°$, equation 28.29 becomes

$$-\frac{2\pi}{3} = \pi \Delta f \lambda u^2 + \frac{1}{2} \pi C_s \lambda^3 u^4 \quad [28.38]$$

Combining equations 28.37 and 28.38 gives a special value for Δf

$$\Delta f_{Sch} = -\left(\frac{4}{3} C_s \lambda\right)^{\frac{1}{2}} \quad [28.39]$$

The subscript denotes the Scherzer defocus value. Since $(1.33)^{1/2} = 1.155$, we have deduced equation 28.35. At this value of Δf we find that we next cross the axis at

$$u_{Sch} = 1.51 C_s^{-\frac{1}{4}} \lambda^{-\frac{3}{4}} \quad [28.40]$$

The resolution at the Scherzer defocus can then be defined as the reciprocal of u_{Sch}

$$r_{Sch} = \frac{1}{1.51} C_s^{\frac{1}{4}} \lambda^{\frac{3}{4}} = 0.66 C_s^{\frac{1}{4}} \lambda^{\frac{3}{4}} \quad [28.41]$$

You will often see this expression with different values for the constant for reasons discussed back in Section 6.6.B (here we are essentially summing the effects of Δf and C_s). The value of the constant can be increased, thus lowering r_{Sch} (i.e., giving higher resolution) if we are less restrictive about the value we choose for χ.

The quantities $(C_s \lambda)^{1/2}$ and $(C_s \lambda^3)^{1/4}$ seen in equations 28.39 and 28.41 are so important in HRTEM that Hawkes (1980) has designated them to be the units 1 Sch and 1 Gl (the scherzer and the glaser) in honor of the two most noted pioneers of HRTEM. Notice that these *units* vary depending on the microscope you're using.

You'll find it interesting to plot the phase shift due to the variation of Δf and C_s using EMS (Section 1.5). An excellent, though advanced, discussion of such diagrams is given by Thon (1975), who describes how they can be used to design phase plates for the TEM. Spence (1988) shows how you can use a plot of nu^{-2} versus u^2 to help you determine experimental values of Δf and C_s; see Figure 30.6A.

28.8. ENVELOPE DAMPING FUNCTIONS

The plots of $\chi(\mathbf{u})$ as a function of $\mathbf{u}$ could extend out as far as you want to plot them. In practice, they don't because of the envelope damping function. In other words, the $\chi(\mathbf{u})$ plot stops where it does because the microscope is incapable of imaging the finest detail due to reasons other than the simple transfer characteristics of a linear system.

We know from Chapters 5 and 6 that resolution is also limited by the spatial coherence of the source and by chromatic effects. We can include these effects in our analysis of images by imposing an envelope function on the transfer function. The result is that higher spatial frequencies that might normally pass through higher-order windows are in fact damped out, as shown in the plot in Figure 28.7.

The exact mathematical form of these envelope functions is complex. In general, the result is described by multiplying the transfer function $T(\mathbf{u})$ by both the chromatic aberration envelope E_c and the spatial coherence envelope E_a to yield an effective transfer function $T_{eff}(\mathbf{u})$

$$T_{eff}(\mathbf{u}) = T(\mathbf{u}) E_c E_a \quad [28.42]$$

The effect of the envelope functions is to impose a virtual aperture in the back focal plane of the objective lens, *regardless* of the setting of focus. If we are going to use a physical aperture to remove unwanted noise, we should make it no larger than the "virtual aperture" present due to this envelope. The presence of this virtual aperture means that higher-order passbands are simply not accessible. This cut-off thus imposes a new resolution limit on the micro-

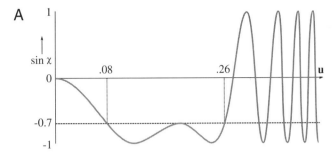

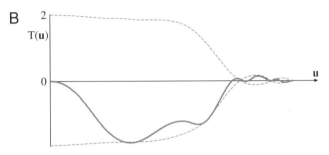

Figure 28.7. (A) $\sin\chi(\mathbf{u})$ versus $\mathbf{u}$ without damping of the higher spatial frequencies. (B) $T(\mathbf{u})$ versus $\mathbf{u}$ modified by the damping envelope (dashed line); $\Delta f = -100$ nm, $C_s = 2.2$ mm.

scope. This is what we earlier called the "information retrieval limit" or simply the "information limit."

If we keep these restrictions in mind then we can say that, up to the instrumental resolution limit, phase-contrast images are directly (i.e., intuitively) interpretable; this limit is set by the crossover at Scherzer defocus or the envelope function, i.e., whichever equals zero first. If the information limit is beyond the Scherzer resolution limit, we need to use image-simulation software (see the next chapter) to interpret any detail beyond the Scherzer limit.

So, you can image columns of atoms along the incident beam direction and their positions are faithfully rendered with respect to one another up to Scherzer resolution. If the microscope is operated at different defocus values, the crossovers in the transfer function make image interpretation more indirect and you have to resort to using computer simulations.

28.9. IMAGING USING PASSBANDS

Because of the focus dependence of the contrast-transfer function, you, the microscope operator, have control over its overall form. For example, the worst case of contrast transfer is where all contrast is minimized. This minimum-contrast (MC) defocus condition (Δf_{MC}) is also known as the dark-field focus condition in STEM imaging and occurs when

$$\sin\chi(\mathbf{u}) \simeq 0.3 \qquad [28.43]$$

or

$$\Delta f_{MC} = -0.44 \left(C_s \lambda\right)^{\frac{1}{2}} \qquad [28.44]$$

The importance of this focus setting is that, when you are actually working on the TEM, you can recognize this focus setting visually on the TEM screen, since it occurs when you can't see anything! If you adjust the focus to this condition visually, you then have a reference point from which you can change to the Scherzer defocus. The procedure is actually quite simple, since you can minimize the contrast easily, providing you have correctly aligned the microscope and corrected the astigmatism.

Some other special settings of the transfer function may also be useful. The idea is to make use of *passbands* or large "windows" in the transfer function to allow higher spatial frequencies to contribute to the image. As you see in Figure 28.8, what this requires is that χ is constant, or $d\chi/du$ small, over a range of u which includes the reflection of interest. These passbands occur periodically with underfocus at values set by

$$\Delta f_p^n = \left\{ \left(\frac{8n+3}{2}\right)(C_s\lambda) \right\}^{\frac{1}{2}} \qquad [28.45]$$

This formula is not an exact relationship but it gives us a good guide; its derivation is given by Spence (1988). The $n = 0$ passband is, in fact, equivalent to the Scherzer defocus setting. This technique gives us access to higher spatial frequencies and thus finer detail in real space. The price we pay is that there are now zeros in the transfer function at lower spatial frequencies. For some applications, the presence of these zeros may be a problem, but for others, useful information can be obtained in these higher passband settings. For a microscope like the JEOL 200 CX, these pass-

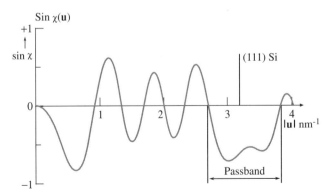

Figure 28.8. Special setting of the transfer function to make use of passbands or "windows" in the transfer function, here optimized to image Si (111).

band settings are −66 nm (Scherzer, or $n = 1$), −129 nm ($n = 2$), −169 nm ($n = 3$), −202 nm ($n = 4$), etc. Note that all are negative values of focus.

We can define an "aberration-free focus" (AFF) condition for any specific crystal (Hashimoto and Endoh 1978). The idea is to set the transfer function so that the gaps will only occur between Bragg reflections. All Bragg reflections would then see a window in the transfer function out to very high order. This aberration-free focus setting is defined by

$$\Delta f_{AFF} \lesseqgtr \left\{2(4m \pm 0.23) + C_s \lambda^3/d^4\right\}\left(d^2/2\lambda\right) \quad [28.46]$$

where $m = 0, 1, 2, 3$, etc. and d is the fundamental lattice spacing of the first-order Bragg beam to be resolved. In the application to a Au crystal in [001] orientation where $d(020) = 0.2035$ nm, using a 100-kV microscope with $C_s = 0.75$ mm, Δf_{AFF} is −53.3 nm. At this setting of focus, the transfer function peaked at −2 for beams 020, 220, 040, 420, 440, and 060.

There is, of course, a catch. We can only use this technique when we know which spatial frequencies we are interested in. In other words, it is great for perfect crystals since we are only concerned with Bragg peaks. If defects are present we lose all the information about the defect, since defects scatter between the Bragg peaks. Any information falling in the gap of the transfer function is lost to the image: in effect, the defect will be invisible!

You should therefore be very cautious in using higher passband settings. You may obtain a pretty picture which does not give a true image of your specimen. If you do use higher-order passbands, you must realize that you are imaging the specimen beyond the instrumental resolution limit so you can't use the intuitive approach for image interpretation. You must know exactly where the zeros are in the transfer function. You can only know that by very careful evaluation of your negatives using diffractograms, computer simulation, and image processing.

28.10. EXPERIMENTAL CONSIDERATIONS

Whenever you are using HRTEM imaging, you must first ask what information you are hoping to obtain. Lattice-fringe images which show lots of straight lines but tell you nothing of where the atoms are located may be just what you need. These fringes are giving you information about the crystal orientation on a very fine scale. Another situation is illustrated by early studies of spinel (Carter *et al.* 1986). You would like to obtain information at, say, 2.3 Å (the spacing of the oxygen 111 planes), but your point-to-point resolution is 2.7 Å. You could still learn a lot about the spinel from the 4.6 Å spinel (111) planes, so you might use an aperture to remove information which only adds uninterpretable detail below 4.6 Å. The difficulty comes when you want to relate your HRTEM image to the atomic structure of your specimen. Then you must remember that all of the above treatment is based upon the TEM specimen behaving as a weak-phase object. Most specimens of interest do *not* satisfy this criterion.

If you look at a typical HRTEM specimen, there will be a wedge-shaped region near the thinnest edge, and thickness extinction contours will be visible. As soon as the first contour is visible, the specimen is already much too thick to behave as a weak-phase object! Multiple scattering limits most phase-contrast imaging conditions for crystalline materials.

> Note that the HRTEM community uses the term "multiple scattering" to denote >1 scattering event. This terminology differs from that used by analytical microscopists, who define "multiple" as >20 scattering events and reserve "plural" for 2–20 events. In HRTEM you never hear of "plural scattering."

Thicker specimens also are susceptible to Fresnel effects associated with spreading of the wave front as it is transmitted through more specimen along the beam direction. Inelastic scattering effects, etc., will also become important as the thickness increases. These effects are not easy to simulate in the computer, although the techniques we will discuss in Chapter 29 are very helpful.

To be really sure that you have correctly interpreted the image, the match between experimental and simulated images should be good over a range of thicknesses and defocus values, as we'll see more clearly in Chapter 30.

We can now summarize the ten steps you need to take to obtain a phase-contrast image with atomic resolution:

- Choose an instrument of low C_s and small λ.
- Align it well; it will take time for the electronic and moving parts to become stable.
- Work with an undersaturated LaB_6 filament and a small condenser aperture (unless you have an FEG; see later).
- Perform current and voltage centering of the objective lens routinely and frequently at high magnification.
- Work in thin, flat, and clean regions of the specimen.

- Orient the specimen using small SAD apertures or bend contours in the image, so the beam is aligned along a zone axis.
- Correct the astigmatism, using optical diffractograms if necessary, but ideally on line (Chapter 30).
- Find the minimum-contrast focus setting and record a through-focus series.
- Record the DP at the same setting of the condenser (calculate α, the convergence semiangle).
- Simulate and/or process the images using available computer codes (Chapter 29).

A comment on alignment: You'll find that it's relatively straightforward to align the electron beam with the current center or voltage center. The result is an image which does not shift as the current changes in the objective lens or the accelerating voltage fluctuates. As you'll appreciate more from Chapter 30, for the highest resolution, it is also critical that the incident beam is precisely parallel to the optic axis of the microscope. If the incident beam is not exactly aligned with the optic axis we can see the coma aberration, which is only important at the highest resolution. (See Hall (1953) for a discussion of other aberrations.) We refer to the process of aligning the beam with the optic axis as "coma-free alignment." The process involves alternately applying equal and opposite beam tilts to the incident beam; you choose the magnitude of the tilt to match the periodicity in the image. If there is a residual beam tilt of the incident beam away from the optic axis, then one image will look more distorted than the other. Adjust the beam-tilt controls until both tilted images look equally distorted. Repeat this procedure for the orthogonal direction. You will need a lot of practice to do this successfully (see Section 30.5).

Some final remarks on experimental techniques: Always remember that specimen orientation is very critical for HRTEM. Always be aware of contamination and damage caused by the electron beam; the specimen will have changed long before you can see the change by eye. Although HRTEM is now so much easier because high-quality video cameras are available to give you an image at TV rates, don't spend any longer than you have to with the beam on the specimen. You must get used to using the TV and the computer. If you are going to do quantitative HRTEM, you'll have to be comfortable with both.

28.11. THE FUTURE FOR HRTEM

The historical approach to HRTEM was: be pleased if you recorded what you saw. Now machines are sufficiently stable that we can reliably record images at different values of Δf. Certainly as important is the availability of computers, as we will discuss in Chapters 29 and 30, since we can "predict" the image for model structures and quantify the contrast of the image. We are thus able to do quantitative HRTEM (QHRTEM or HRQTEM!).

Another approach to improve resolution is provided by the FEGTEM. Such a beam is now highly coherent, so the envelope function shown in Figure 28.7 extends to greater values of **u**. The computer now becomes indispensable because we have to interpret images which have contrast reversals beyond Scherzer defocus. If a carefully designed multipole lens is inserted into an HRTEM, Rose (1990, 1991) has shown that it is, in principle, possible to correct C_s!

When this happens we will have to rethink our approach to HRTEM. Think what will happen to the scherzer

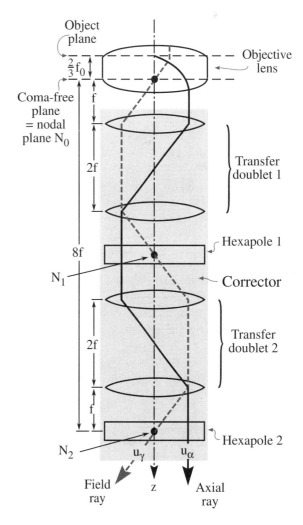

Figure 28.9. The post-objective lens corrector system proposed by Rose to correct spherical aberrations in the objective lens.

and the glaser. What will the Gl/Sch be? The corrector proposed by Rose is shown schematically in Figure 28.9. It is a combination of round lenses and hexapoles, all of which are magnetic elements. The hexapoles don't affect the paraxial path of the rays and only need to be stabilized to an accuracy of 1 in 10^4 to give atomic resolution. When C_s is zero, the specimen resolution limit will be determined by C_c

$$d_{C_s = 0} \simeq \left[\left(\frac{\Delta E}{E} \right) \lambda C_c \right]^{\frac{1}{2}} \quad [28.47]$$

If C_c = 2 mm and $\Delta E \sim 0.3$ eV, a 200-kV FEGTEM could achieve a resolution of 0.8 Å. If C_c is also corrected, which we'll see how to do in Chapter 40, it is possible that the resolution will become limited by the fifth-order spherical aberration constant. In practice, it will be important to correct C_c in the lens design first. In Rose's proposed lens, C_s = 3 mm and the resolution limit is 0.28 Å for this 200-kV FEGTEM. Other lens defects will limit this to ~0.5 Å, but with a price of $12M for a 1.25-MeV machine which damages your specimen in seconds, the Rose corrector could be quite important!

28.12. THE TEM AS A LINEAR SYSTEM

The discussion we went through above is an example of a much larger topic known as information theory (Shannon and Weaver 1964, Van Dyck 1992). The concept of a "phase-contrast transfer function" is central to this field. So you can understand the practice of phase-contrast imaging at high resolution, we will briefly discuss the way an information specialist might view this process. We will define the transfer function in elementary terms, and make detailed reference to phase-contrast imaging in the TEM.

Remember, the purpose of the TEM is to transmit information about the specimen to the image. We can thus consider the microscope to be an "information channel" and use the concepts of information theory:

- The input signal comes from the specimen.
- The output signal is the image.

If we neglect the effects of noise, there is a unique relation between the input signal and the output signal, determined by the optical system of the microscope.

Most information theory treats linear systems. A linear system is one which is characterized by the property that if

$$S_0(r_0) \rightarrow \text{Transmission System} \rightarrow S_1(r_1),$$
and if
$$S'_0(r_0) \rightarrow \text{Transmission System} \rightarrow S'_1(r_1),$$

(the prime here denotes the derivative), then the system is linear if

$$a(S_0) + b(S'_0) \rightarrow \text{Transmission System} \rightarrow a(S_1) + b(S'_1)$$

for any values of a and b.

The linear relation between input and output signals can be described by the concept of the transfer function. Overall, the transfer function relates an input spectrum to an output spectrum, and it operates only in the frequency domain.

In general, for a linear system, if we know the transfer function, then the relation between S_0 and S_1 is uniquely defined. On the other hand, if the relation between S_0 and S_1 could be empirically determined, then we can deduce the transfer function.

One of the best examples of a linear system is an electrical transmission cable. The transfer of electrical signals through transmission lines can be made linear enough for the above theory to apply. Conversely, the transfer of mass-thickness information from a specimen to the optical density of a developed photographic negative is far from linear, and the above theory does not apply. Then why should we bother to discuss this in HRTEM? The answer lies in finding an appropriate linear relation between the object and the image.

> Schrödinger's wave equation is linear. Therefore, the amplitudes of an electron wave in the specimen are linearly related to the amplitudes of an electron wave in the image.

28.13. FEGTEMs AND THE INFORMATION LIMIT

We've mentioned that an FEG reduces the instrumental contribution to chromatic aberrations and extends the envelope function to larger values of **u**. This means that information with higher spatial frequencies is transferred to the image. We've just analyzed the Scherzer defocus problem, so now we'll consider the information limit. The reason for emphasizing the FEG here is that it really makes a difference and we're just beginning to learn how to use this information: the contrast reversals mean that any image interpretation is not intuitive. The topic has been laid out in two papers (Van Dyck and de Jong 1992, de Jong and Van Dyck 1993).

Since the information limit is determined by the envelope function, this is split into its separate terms. The total envelope function, $E_T(\mathbf{u})$, is the product of all of these

$$E_T(\mathbf{u}) = E_c(\mathbf{u})E_s(\mathbf{u})E_d(\mathbf{u})E_v(\mathbf{u})E_D(\mathbf{u}) \quad [28.48]$$

28 ■ HIGH-RESOLUTION TEM

The individual envelope functions in equation 28.48 are:

$E_c(\mathbf{u})$: for chromatic aberration.
$E_s(\mathbf{u})$: for the source dependence due to the small spread of angles from the probe.
$E_d(\mathbf{u})$: for specimen drift.
$E_v(\mathbf{u})$: for specimen vibration.
$E_D(\mathbf{u})$: for the detector.

As you can see, some of these envelope functions are new, some are old. We won't discuss all the functions; we'll only mention a couple of the key points.

The chromatic aberration is well known, and its envelope function $E_c(\mathbf{u})$ can be expressed by the equation

$$E_c(\mathbf{u}) = \exp\left[-\frac{1}{2}(\pi\lambda\delta)^2 u^4\right] \quad [28.49]$$

where c reminds us that this is a chromatic aberration and δ is the defocus spread due to this aberration

$$\delta = C_c\left[4\left(\frac{\Delta I_{obj}}{I_{obj}}\right)^2 + \left(\frac{\Delta E}{V_{acc}}\right)^2 + \left(\frac{\Delta V_{acc}}{V_{acc}}\right)^2\right]^{\frac{1}{2}} \quad [28.50]$$

The terms $\Delta V_{acc}/V_{acc}$ and $\Delta I_{obj}/I_{obj}$ are the instabilities in the high-voltage supply and the objective-lens current. $\Delta E/V_{acc}$ is the intrinsic energy spread in the electron gun. Notice that ΔE and ΔV are different: ΔV depends on how well we can control the voltage supply whereas ΔE depends on our choice of electron source (see Chapter 5). If we neglect any other contributions to the envelope function, then we can define an information limit due to instrument chromatic aberrations by ρ_c

$$\rho_c = \left(\frac{\pi\lambda\delta}{\sqrt{2\ln(s)}}\right)^{\frac{1}{2}} \quad [28.51]$$

where e^{-s} is the cut-off value for the envelope. If we take $\ln_e s$ to be 2, then

$$\rho_c = \left(\frac{\pi\lambda\delta}{2}\right)^{\frac{1}{2}} \quad [28.52]$$

The source-dependent envelope function is new because, until we have an FEG, we don't usually consider the "source of a probe." If we imagine that the source has a Gaussian distribution, we have an envelope function $E_s(\mathbf{u})$ given by

$$E_s(\mathbf{u}) = \exp\left[\left(\frac{\pi\alpha}{\lambda}\right)^2\left(\frac{\partial\chi(\mathbf{u})}{\partial u}\right)^2\right]$$

$$= \exp\left[-\left(\frac{\pi\alpha}{\lambda}\right)^2\left(C_s\lambda^3 u^3 + \Delta f\lambda u\right)^2\right] \quad [28.53]$$

Here, α is the semiangle characterizing the Gaussian distribution. What this equation tells us is that if α is too large (≥ 1 mrad) it can limit the information limit. If we say that u must lie between u and some maximum value u_{max}, we can maximize the argument of the exponential in equation 28.53 to give an optimum focus

$$\Delta f_{opt} = -\frac{3}{4}C_s\lambda^2 u_{max}^2 = -\frac{3}{4}\frac{C_s\lambda^2}{\rho_i^2} \quad [28.54]$$

In this equation, ρ_i is the information limit of the microscope (which is how we chose u_{max}). This defocus value will be important later when we discuss holography in an FEGTEM. The two curves shown in Figure 28.10 illustrate how this envelope function varies within Δf. It can also be optimized by decreasing the semiangle α. With a little more manipulation, de Jong and Van Dyck (1993) show that the information limit due to the limited coherence of the source is given by

$$\rho_\alpha = \left(\frac{6\pi\alpha a}{\lambda\sqrt{\ln(s)}}\rho_s^4\right)^{\frac{1}{3}} \quad [28.55]$$

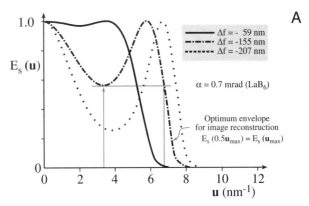

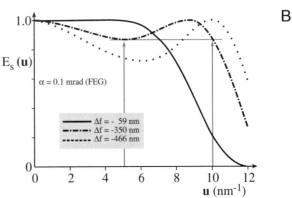

Figure 28.10. Variations in the envelope function, $E_s(\mathbf{u})$, for different objective lens defocus: (A) LaB$_6$ source, (B) FEG.

The envelope functions for the drift and vibration represent a new method for taking account of these two unavoidable quantities. We'll just quote the results for the two "information limits" which are the crossover values for the two envelope functions $E_d(\mathbf{u})$ and $E_v(\mathbf{u})$

$$\rho_d = \frac{\pi d}{\sqrt{6 \ln(s)}} \qquad [28.56]$$

and

$$\rho_v = \frac{\pi v}{\sqrt{\ln(s)}} \qquad [28.57]$$

In these equations, d is the total drift during the exposure time, t_{exp}, so $d = v_d t_{exp}$ for a drift velocity v_d; v is the amplitude of the vibration.

The detector envelope function, $E_D(\mathbf{u})$, is something we never worried about with film, but CCD cameras have a limited number of pixels, i.e., we only have a limited number of resolved image points. This envelope function results from two effects:

- Delocalization of the information in the image.
- The finite pixel size.

The idea is simple but the math is more difficult. If the image which is actually captured by the camera is a circle, we can say that if R is less than R_w, the radius of the window, then we capture the information; if it's greater than R_w, we don't. So the CCD detector is acting like an aperture! Now de Jong and Van Dyck show how u_{max} is related to R_w

$$\alpha C_s \lambda^3 u_{max}^3 = R_w \qquad [28.58]$$

The important result is that the delocalization of the information in the image must be less than the half-width of the CCD detector array.

> Image delocalization depends on **u**. It is large when $\partial \chi(\mathbf{u})/\partial u$ oscillates rapidly, as it does for large **u**, i.e., where we are placing the information limit.

The value of R_w is related to the number of pixels, N, and their size, D

$$R_w = \tfrac{1}{2} N D \qquad [28.59]$$

The information limit due to the detector (i.e., the crossover values of the detector envelope functions $E_D(\mathbf{u})$) is

$$\rho_D = \left(\frac{12\sqrt{2}\pi a}{N\sqrt{\ln(s)}} \right)^{\frac{1}{4}} \rho_s \qquad [28.60]$$

Clearly, we can decrease ρ_D by increasing N, but not quickly. With this analysis in mind we can summarize the conditions necessary for ρ_i to be limited by chromatic aberration

$$\alpha \leq \frac{\lambda}{6\pi a \rho_s} \left(\frac{\rho_c}{\rho_s} \right)^3 \qquad [28.61]$$

$$N \geq 12\sqrt{2}\pi a \left(\frac{\rho_s}{\rho_c} \right)^4 \qquad [28.62]$$

$$d \leq \frac{\sqrt{6}}{\pi} \rho_c \simeq 0.8 \rho_c \qquad [28.63]$$

$$u \leq \frac{1}{\pi} \rho_c \simeq 0.3 \rho_c \qquad [28.64]$$

Table 28.1 gives some numerical examples of what these equations mean.

To see whether we will ever reach the information limit, we have to consider the effect of the noise. We know that the signal-to-noise ratio is proportional to $\beta^{1/2}$, where β is the brightness of the electron gun. If the smallest image element we need to examine has an area ρ_i^2, then the background signal I_0 is given by

$$I_0 = D\rho_i^2 = \beta\pi\alpha^2 t \rho_i^2 \qquad [28.65]$$

where D is the electron dose, α is the semiangle of convergence, and t is the time.

For white noise, the noise in an element will be related to $I_0^{1/2}$. The total contrast in our small pixel can be written as $DEF\rho_i^2$, where D is the dose, E is the envelope function, F is the structure factor, and ρ_i^2 is the area of the pixel. Now we can say that the minimum detectable signal-to-noise ratio is k, which gives

$$DEF\rho_i^2 = kI_0^{\frac{1}{2}} = k\rho_i D^{\frac{1}{2}} \qquad [28.66]$$

Table 28.1. Maximum Convergence Semiangle α and Minimum Number of Unusable Image Points N for Different Values of the Point Resolution to (Chromatic) Aberration Limit Ratio (ρ_s / ρ_c).[a]

	α (mrad)		N (pix)	
ρ_s / ρ_c	ε_0	ε_{opt}	ε_0	ε_{opt}
1	0.58	2.3	53	13
1.5	0.17	0.69	270	67
2	0.07	0.30	853	213
2.5	0.04	0.15	2082	521
3	0.02	0.09	4320	1080

[a]ε_0: Gaussian focus, ε_{opt}: optimum focus; $\lambda = 0.011 \rho_s$ (de Jong and Van Dyck 1993).

Therefore, the signal-to-noise ratio for $\alpha = 1$ mrad and $t = 1$ second can be written as

$$s_0 = 443 \, \rho_i \frac{F}{k} \beta^{\frac{1}{2}} \quad [28.67]$$

Now you can use some real numbers: take $k = 2$ (think what this means for the minimum contrast), and assume that β is 10^{10} Am^{-2}sr^{-1} for a LaB$_6$ gun and 10^{13} Am^{-2}sr^{-1} for a Schottky FEG. You can show that for $\rho_i = 0.15$ nm (a LaB$_6$ gun), $\ln_e s_0$ is 1.2 to 2.2, whereas for an FEG, $\rho_i = 0.1$ nm and $\ln_e s_0$ is 4.5 to 5.2. You can also appreciate why s_0 depends on your material: low atomic numbers mean weak scattering. For our last two equations we'll again quote de Jong and Van Dyck. We can deduce optimum values for both the angle of convergence α and the exposure t by differentiating the envelope equations

$$\alpha_{opt} = \frac{1}{k_s\sqrt{2}}\left(\frac{\rho_i}{\rho_c}\right)^3 = \frac{1}{6\pi a\sqrt{2}} \frac{\lambda}{\rho_s}\left(\frac{\rho_i}{\rho_s}\right)^3 \quad [28.68]$$

$$t_{opt} = \frac{1}{2k_d}\left(\frac{\rho_i}{\rho_c}\right) \approx 0.39\left(\frac{\rho_i}{v_d}\right) \quad [28.69]$$

Notice that α_{opt} depends not only on ρ_s and ρ_i, but also on λ (of course, ρ_s and ρ_i also depend on λ), and that t_{opt} only depends on the drift rate; fortunately v_d will never be zero!

We can now summarize these new concepts:

- Microscopy is much more complex when you try to use the information limit rather than the Scherzer limit!
- If you want to use a computer, the size of the CCD camera will also affect the actual information limit; this is the effect of $E_D(\mathbf{u})$.
- Drift and vibrations must be minimized or they will determine your resolution; these contributions were described by $E_d(\mathbf{u})$ and $E_v(\mathbf{u})$.
- When everything else is perfect, your resolution will be controlled by the signal-to-noise ratio of the detector and the coherence functions, $E_D(\mathbf{u})$ and $E_c(\mathbf{u})$.
- An FEGTEM improves the information limit because of the large increase in the brightness, β. This increase allows us to decrease α, increase the dose, and increase the signal-to-noise ratio.

28.14. SOME DIFFICULTIES IN USING AN FEG

We've discussed the advantages of using an FEG for HRTEM, but there are some practical difficulties which have been analyzed by Otten and Coene (1993). A cold FEG (CFEG) allows us to extract a very high current per unit area, but the total area of the emitting region is very small so that the extraction current is <5 nA. This current can be increased if we thermally assist the field emission by heating the Schottky emitter to ~1500°C. It gives the same high brightness, but a larger maximum current because of the larger emitting area. So what are the difficulties?

- The emitter area may be so small that we have to "fan" the beam in order to illuminate the area used in TEM. This fanning may actually increase the effect of coma aberration (a radial aberration as noted in Section 28.10). If a CFEG has a source size of ~3 nm, we can study ~15 nm with a 5× magnification. The Schottky source has a diameter ~10× greater, and the price you pay for this is a decrease in spatial coherence, and a larger energy spread.
- Correcting astigmatism is very tricky with an FEG. As you can appreciate from Figure 28.11, if the image is astigmatic you'll see this at all defocus settings with an LaB$_6$ source. In an FEG, when astigmatism is present, all the images look similar and you can't use the technique of finding the minimum-contrast defocus (at ~0.4 sch) to define Δf. If you try to use the wobbler to do coma-free alignment (Otten and Coene 1993), that fails too because you can't interpret the focus difference between two FEG images for the two wobbler directions. There is a solution to finding Δf, fortunately; either use on-line processing (Chapter 30) or converge the beam! The latter way deteriorates the spatial coherence and you've made your $1.5M FEG machine behave like an old $200k LaB$_6$ machine.
- Through-focus series are a challenge, because you can now use a very large range of Δf values, and it becomes a major task just to determine your value of Δf.
- Image delocalization occurs when detail in the image is displaced relative to its "true" location in the specimen. The effect is emphasized by the graph shown in Figure 28.12 and becomes worse as you go away from Scherzer defocus. The effect is illustrated in Figure 28.13, where fringes from the gold particles can appear outside the particle. If we rewrite equation 28.30, we can express the delocalization as

$$\Delta R = \lambda u \left(\Delta f + C_s \lambda^2 u^2\right) \quad [28.70]$$

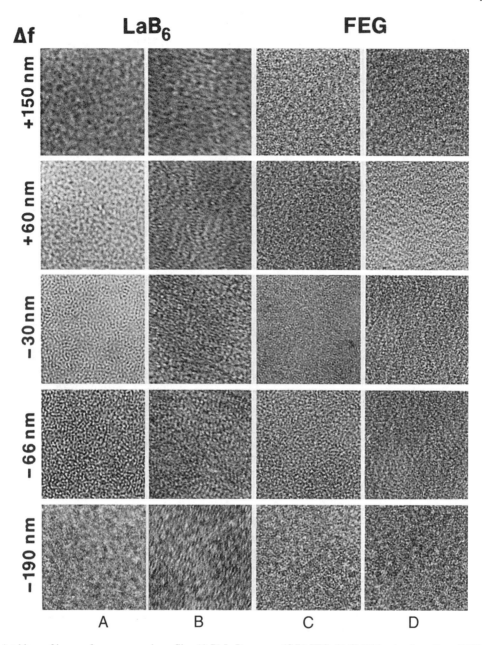

Figure 28.11. A tableau of images from an amorphous film. (A,B) LaB$_6$ source; (C,D) FEG. (A,C) Without astigmatism, (B,D) with astigmatism. With LaB$_6$ you can easily see the astigmatism while with an FEG, you can't.

You may notice a similarity between this equation and that for the SAD error (Chapter 11). Two values have been proposed for Δf_{opt}, the optimum defocus setting to minimize delocalization (Coene and Janssen 1992, Lichte 1991). They give an optimum value for the defocus of

$$\Delta f_{opt} = -MC_s\lambda^2 u_{max}^2 \quad [28.71]$$

where M is a factor between 0.75 and 1. A value for ΔR_{min} is close to

$$\Delta R_{min} = \frac{1}{4}C_s\lambda^3 u_{max}^3 \quad [28.72]$$

The actual value of M is determined by where you define the cut-off value for **u**. There are three conclusions on delocalization:

28 ■ HIGH-RESOLUTION TEM

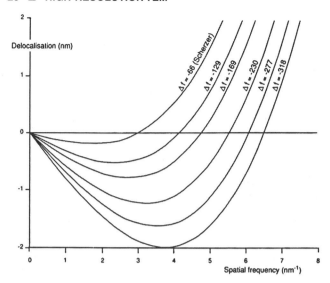

Figure 28.12. Image delocalization plotted as Δf is changed for a Philips CM20 FEG with $C_s = 1.2$ mm.

- As C_s decreases, delocalization decreases.
- As λ decreases (accelerating voltage increases), delocalization decreases.
- Delocalization cannot be avoided in an FEG!

28.15. SELECTIVELY IMAGING SUBLATTICES

In Chapter 16, where we discussed the ordered intermetallic alloys, we saw that many materials with a large-unit cell are closely related to a material with a smaller unit cell. If the two structures don't have the same symmetry, then the two unit cells can show several different orientation relationships, as was the case for vanadium carbide.

We can use this information to form different high-resolution images instead of different DF images. Two [001] DPs from an ordered alloy of Au_4Mn are shown in Figure 28.14 together with a schematic of one pattern (Amelinckx et al. 1993). Two domains are present in the combined pattern. Both patterns have 4-fold symmetry, but they are rotated relative to one another. If we use the DF lattice-imaging mode and exclude all the fcc reflections using the objective aperture, we form an image like that shown in Figure 28.15. The two variants are not only easily recognized, but we know where they are with an accuracy of atomic dimensions. If you're used to grain boundary theory, the original cell has become the coincident-site lattice (CSL) in reciprocal space and the two sublattices are like grains with a small Σ. This approach has been used to

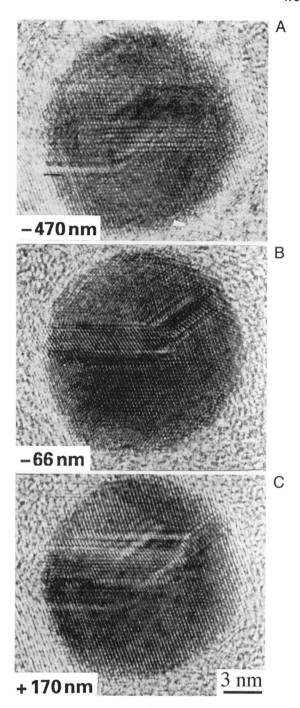

Figure 28.13. Experimental images showing delocalization in HRTEM images of a Au particle: (A) underfocus, (B) Scherzer defocus, (C) overfocus.

estimate the size of very small particles of $NiFe_2O_4$ spinel which are completely contained in a matrix of NiO, as illustrated in Figure 28.16 (Rasmussen et al. 1995). The lattice parameter of the spinel is twice that of the NiO but the latter is generally above and below the particle. This ap-

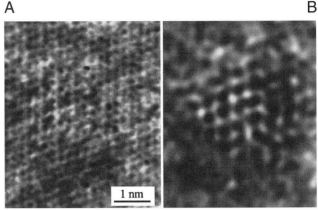

Figure 28.16. (A) The experimental image of a small spinel particle in a NiO matrix. The NiO is thicker and dominates the image. (B) After filtering out the NiO contribution (its lattice parameter is twice that of the spinel) we can see the spinel particle and estimate its size. See also Figures 30.2 and 30.3.

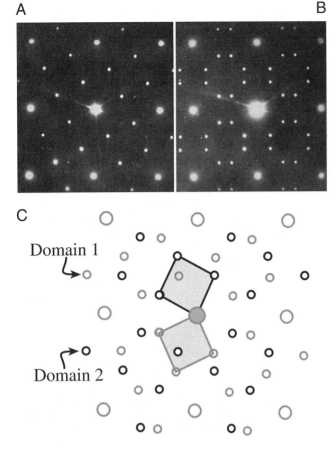

Figure 28.14. Two [001] DPs from an ordered alloy of Au_4Mn. (A) One domain. (B) Two symmetry-related domains. (C) Schematic diagram showing how (B) arises from the relative rotation of the two domains.

proach is therefore quite difficult, especially if, as in the analysis of Figure 28.16, the shape of the particle is important. Then you need to resort to simulation and processing, as we'll discuss in Chapter 29.

28.16. INTERFACES AND SURFACES

Interfaces of all kinds have been extensively studied by HRTEM, in part because we need the near-atomic resolution, but sometimes it's because they make ideal subjects for study! Point defects require extensive image processing and simulation, dislocations move, but interfaces seem to remain stationary if you're careful. However, we are always limited as to which interfaces we can study.

> The fundamental requirement is that the interface plane *must* be parallel to the electron beam.

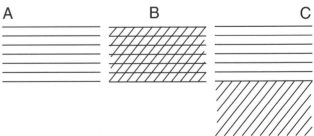

Figure 28.17. Schematic HRTEM images of grain boundaries showing (A) one set of fringes in one grain; (B) specimen tilted to give crossing fringes in one grain; (C) one set of fringes in each grain. In each case the boundary remains parallel to the beam.

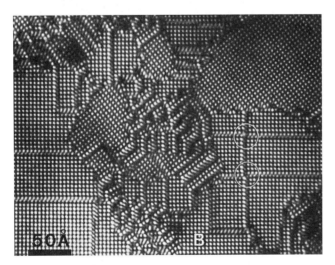

Figure 28.15. DF lattice image of Au_4Mn using an objective diaphragm to exclude all the fcc reflections. The two differently oriented domains correspond to the two orientations in Figure 28.14C.

If a low-index plane in one grain (but preferably in both grains) is parallel to the interface, you're in business. The problem is that you can rarely be sure that this is the case, but because you are looking at a very thin specimen, the projected width of even a tilted boundary is small. So you can tilt the specimen to look down a pole in one grain or to make the beam parallel to a low-index plane in the second grain as shown in Figure 28.17.

If you are lucky (and we often are because we only study tilt boundaries by HRTEM) you can produce crossed fringes in both grains. A selection of images is shown in Figure 28.18. Here you can see structured boundaries, boundaries with an amorphous layer between the grains, interfaces between two different materials, and a surface profile image. We can make some general comments about these images:

- Even a "low"-resolution, lattice-fringe image gives you information on the local topology of your interface.
- If the layer of amorphous material in the boundary is quite thick (> 5 nm), you can see it directly.
- You can quite easily see detail like five-membered rings in grain boundaries, but you should be wary of interpretation until you've covered Chapter 29.
- You can see abrupt interfaces at near-atomic dimensions.

Now we can also list our concerns:

- Has grooving at the interface affected the appearance of the image? The answer is "maybe," but does it affect what you wanted to learn?
- Is the phase boundary as chemically abrupt as it is structurally? It is very difficult to answer this. The appearance of the image changes at the interface in Figure 28.18C mainly because the total number and location of the cations (Fe^{3+} and Ni^{2+}) changes, not because there is a 2:1 ratio of Fe:Ni.
- Are all of the black spots in Figure 28.18D complete columns? Clearly this question is related to the first.

We will address some of these problems in Chapter 29 but we can make some comments now:

- The quality of your imaging data will be governed by how well you prepare your specimen. Nearly all subsequent analysis will assume that it has a uniform thickness across the interface.

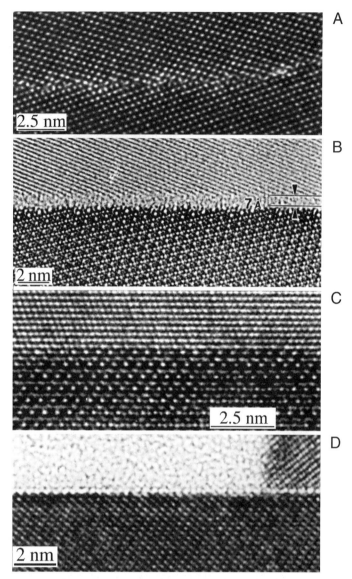

Figure 28.18. Examples of HRTEM images of planar interfaces. (A) Grain boundary in Ge; (B) grain boundary in Si_3N_4 with a layer of glass along the interface; (C) phase boundary separating NiO and $NiAl_2O_4$; (D) profile image of the (0001) surface of Fe_2O_3.

  If you do not *know* that this is so, then your interpretation may be questioned.
- Crystalline grains also thin at different rates if they have different orientations, or different structures, or different chemistries. The grain boundary layer, whether crystalline or amorphous, will also thin at a different rate. Why? Because the bonding and density are different. So careful specimen preparation is absolutely critical.

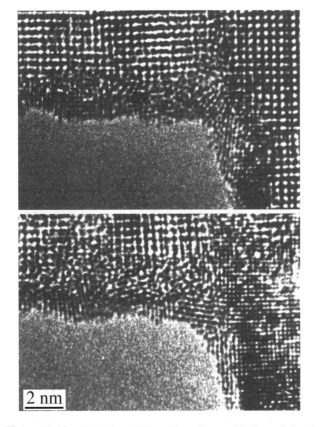

Figure 28.19. Reduction of Nb oxide to the metal by beam-induced loss of oxygen during observation of the edge of the foil. The reduction increases with time (lower image).

- You can learn a lot about your interface using HRTEM without trying to use atomic-resolution imaging.
- The longer you look at your specimen, the more it will differ from what you started with. Use at least a pseudo-low-dose approach, if possible. Figure 28.19 illustrates an extreme example. Here, the oxide has been completely reduced to the metal at the edge of the foil. Of course, this now provides a method for studying the reduction of oxides under the electron beam in the presence of hydrocarbons and a good vacuum!

28.17. INCOMMENSURATE STRUCTURES

We'll illustrate this topic by considering several types of incommensurate (modulated) structures. In each case, the structure consists of a "parent" structure to which we then add a periodic modulation by means of an internal planar defect. Van Landuyt et al. (1991) have characterized three different types of incommensurate structures:

- Periodic modules of the parent separated by interfaces. The interface may be a stacking fault (SF), twin boundary (TB), anti-phase boundary (APB), inversion domain boundary (IDB), crystallographic shear (CS) plane, or discommensuration wall.
- A parent structure with a superimposed periodic deformation wave with a larger periodicity.
- A parent structure where the composition or site occupancy changes periodically.

The next complication is that we can find commensurate and incommensurate structures, and also structures where the modulation is variable. To understand how a structure can be incommensurate consider Figure 28.20, where we've placed a planar defect after every seventh layer so that it expands the lattice by δ every seventh plane. The parent lattice will show a spot spacing in the DP proportional to d^{-1} but the "superlattice" will have a periodicity of Δ^{-1}, which means that we need not have a simple relationship between the two arrays of spots.

These different kinds of modulation can be combined! We'll illustrate this type of specimen with two examples. The Bi-Sr-Ca-Cu-O superconductor provides a good illustration of this type of structure. The parent is a perovskite-like cube; we may have two, three, or four layers of the perovskite with each group separated from the next by a bismuth oxide layer. The formula can be written as $Bi_2Sr_2Ca_nCu_{n+1}O_{2n+\delta}$, so for $n = 1$ we have a sequence of layers (planes) described as BiO-SrO-CuO_2-Ca-CuO_2-SrO-BiO, as shown in Figure 28.21. The DPs depend on the particular value of n in the chemical formula and show rows of satellite reflections due to the modulation of the

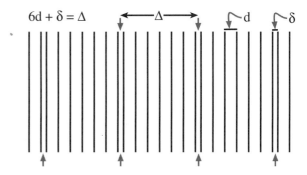

Figure 28.20. An incommensurate structure formed by inserting a planar defect after every seventh layer to expand the lattice by δ every seventh plane.

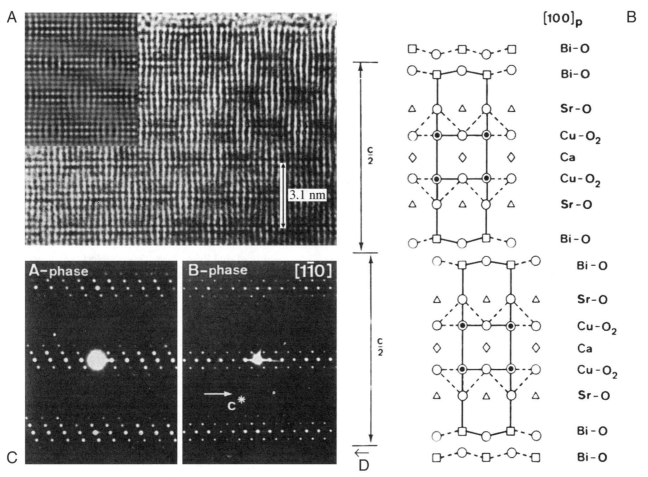

Figure 28.21. (A) HRTEM image of the superconductor $Bi_2Sr_2Ca_nCu_{n+1}O_{2n+\delta}$; the simulation (inset) assumes that the lattice relaxation occurs in the Bi-O layer. (B) For $n = 1$, the structure is built up from blocks which are shifted relative to one another. The DPs are from (C) the n = o phase and (D) the $n = 1.2$ phase. Notice that the spacings of the spots (the satellite sequence) are different.

basic structure. When we form the HRTEM image, the lattice planes appear wavy although we can recognize an orthorhombic pattern. The wavy modulation in the image is probably due to excess oxygen in the BiO layers: the BiO layers don't fit very well to the perovskite block, but the misfit stresses can be relaxed by introducing excess oxygen. There are several clear lessons from this example:

- HRTEM is essential if you are to understand such structures.
- You need supplementary information, such as the chemistry of the specimen.
- Images of this structure produced with the beam along the orthogonal direction would be difficult to explain.

Modulated structures are not confined to the superconductors. In fact such structures are ubiquitous in the materials world. Many useful engineering alloys exhibit spinodal de-

composition and the spinodal wavelength can be directly measured from the extra spots in the DP (Butler and Thomas, 1970). Many ceramics are described as polytypes, e.g., SiC, or polytypoids, which are just polytypes with larger composition fluctuations from layer to layer (e.g. SiAlONs) (Bailey 1977). These structures consist of random or locally ordered stacking of specific atomic layers, which often give predictable effects in the DP. All these materials are particularly amenable to HRTEM analysis, because you can image the individual modulations, but still characterize the overall structure with conventional amplitude contrast.

28.18. QUASICRYSTALS

The study of quasicrystals continues to be a challenge for HRTEM, since these materials do not have the transla-

Figure 28.22. Tenfold symmetry in a decagonal Al-Mn-Pd quasicrystal.

tional symmetry which we associate with crystals. However, they are strongly ordered, as you can appreciate from Figure 28.22 (Nissen and Beeli 1991). The HRTEM image shows many sharp white spots from a stable decagonal quasicrystal of Al-Mn-Pd. The DP from another specimen also showed very strong, clear, well-defined spots. In our discussion of DPs we associated each spot with a single set of planes, which were present throughout the specimen. Although the quasicrystals do not contain such planes, there is clearly far more order than in an amorphous material. You can indeed see that the spots in the HRTEM image are aligned in certain well-defined directions, but the spacing is difficult to identify. We have a growing understanding of these materials and it appears that the spots in the HRTEM image are this sharp because, at least in decagonal quasicrystals (but not in icosahedral ones), they really do correspond to columns; we don't need translational periodicity along the column. In fact, we can rotate the quasicrystal (they can be grown as large as 1 mm) to reveal twofold and threefold axes, as illustrated in Figure 28.23. You can see how this might arise by looking along the rows of spots in Figure 28.22. We can draw some interesting lessons from the use of TEM to study quasicrystals:

- HRTEM excels when materials are ordered on a local scale.
- For HRTEM, we need the atoms to align in columns because this is a "projection technique," but the distribution along the column is not so critical.
- SAD and HRTEM should be used in a complementary fashion.

28.19. SINGLE ATOMS

You have read that it has become possible to study materials at atomic resolution in the TEM quite recently. So you may be surprised to find that many groups have been reporting studies of individual atoms since about 1970! Reimer (1993) gives over 20 references to this topic. The techniques used include phase contrast and amplitude contrast in a conventional TEM, and (see Section 22.4) a dedicated STEM. Parsons *et al.* (1973) used mellitic acid molecules stained with uranyl ions from uranyl acetate so the atoms that were imaged are heavy. Parsons *et al.* knew that the uranium atoms would be 1 nm apart at each apex of an equilateral triangle and they knew that there were 10^{13} of these per cm^2 supported on a thin (0.8 nm) film of evaporated carbon. The challenge is recognizing that the contrast from the individual uranium atoms reverses as you change defocus, just as we've seen for columns of atoms. You can see this effect in Figure 28.24.

Some points to notice:

- This is a case where we really do have "white atoms or black atoms"!

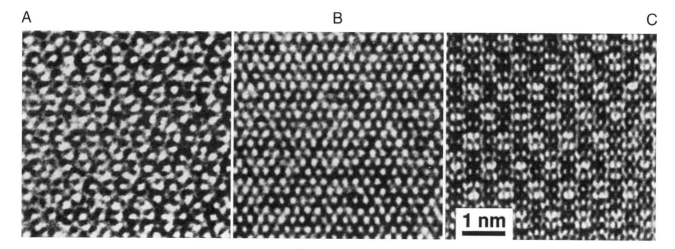

Figure 28.23. Images of (A) fivefold, (B) threefold and (C) twofold projections of an Al-Li-Cu quasicrystal; $\Delta f = -27$ nm.

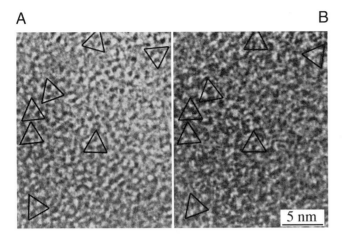

Figure 28.24. Images from triangular arrangements of uranium atoms at different values of defocus.

- Parsons *et al.* (1973) used a Siemens 101 TEM operating at 100 kV with a point-to-point resolution of about 0.33 nm; this is not today's state-of-the-art machine!
- The specimen was so stable that they could do "through-focus" imaging.

Z-contrast imaging heralded the arrival of the STEM as a real research tool: atoms were seen to move on the surface, agglomerate, etc. The imaging mode was essentially high-angle dark-field so that the heavy atoms are by far the strongest scatterers and appear bright, as we discussed in Chapter 22. The difficulty with this approach is that it requires an FEG, whereas almost any TEM operating today can produce images like those first demonstrated by Parsons *et al.*

CHAPTER SUMMARY

The major problem that separates this chapter from Chapter 27 is the language. In HRTEM, the language is that of physics or electrical engineering; you can get so involved with the language and the equations that you miss the point. Having said that, you must know the following terms and understand what they mean or don't mean:

- Point-spread function.
- Contrast transfer function (CTF).
- Weak-phase-object approximation (WPOA).

With this understanding of the restrictions that our model involves, we can now consider the simulation of high-resolution images. If you want to delve further into the theory, we recommend starting with the works by Cowley (1981) and Spence (1988), but you will need to have a strong background in math and physics to appreciate fully the further subtleties of more complex models. Always keep in mind that all of the above discussion was concerned with arriving at models or approximations:

- We model the effect of the lens.
- Then we model the effect of the specimen.
- Finally, we combine the two models.

We will take the next two chapters to achieve these three tasks. Once you pass those chapters you are ready to attack John Spence's text.

REFERENCES

General References

Buseck, P.R., Cowley, J.M., and Eyring, L. (Eds.) (1988) *High-Resolution Electron Microscopy and Associated Techniques*, Oxford University Press, New York.

Hawkes, P.W. (1980) *Ultramicroscopy* **5**, 67. An enjoyable and informative diversion.

Horiuchi, S. (1994) *Fundamentals of High-Resolution Transmission Electron Microscopy*, North-Holland, Amsterdam.

Spence, J.C.H. (1988) *Experimental High-Resolution Electron Microscopy*, 2nd edition, Oxford University Press, Oxford, United Kingdom.

Vladár, A.V., Postek, M.T., and Davilla, C.D. (1995) *Scanning* **17**, 287. Looks at the SEM as a hi-fi instrument.

Specific References

Amelinckx, S., Milat, O., and Van Tendeloo, G. (1993) *Ultramicroscopy* **51**, 90.

Bailey, S.J. (1977) International Union of Crystallography, *Acta Cryst.* **A33**, 681.

Butler, E.P. and Thomas, G. (1970) *Acta Met.* **18**, 347.

Carter, C.B., Elgat, Z., and Shaw, T.M. (1986) *Phil. Mag.* **A55**, 1.

Coene, W. and Janssen, A.J.E.M. (1992) in *Signal and Image Processing in Microscopy and Microanalysis,* Scanning Microscopy Supplement **6** (Ed. P.W. Hawkes) p. 379, SEM Inc. AMF, O'Hare, Illinois.

Cowley, J.M. (1981) *Diffraction Physics,* 2nd edition, North-Holland, Amsterdam.

Cowley, J.M. (1992) in *Electron Diffraction Techniques.* **1** (Ed. J.M. Cowley) p. 1, I.U.Cr., Oxford Science Publication.

de Jong, A.F. and Van Dyck, D. (1993) *Ultramicroscopy* **49**, 66.

Fejes, P.L. (1977) *Acta Cryst.* **A33**, 109.

Hall, C.E. (1953) *Introduction to Electron Microscopy,* McGraw-Hill, New York.

Hashimoto, H. and Endoh, H. (1978) *Electron Diffraction 1927–1977* (Eds. P.J. Dobson, J.B. Pendry, and C.J. Humphreys), p. 188, The Institute of Physics, Bristol and London.

Lichte, H. (1991) *Ultramicroscopy* **38**, 13.

Nissen H.-U. and Beeli, C. (1991) in *High Resolution Electron Microscopy: Fundamentals and Applications* (Eds. J. Heydenreich and W. Neumann), p. 272, Institut für Festkörperphysik und Electronmikroskopie, Halle/Saale, Germany.

Otten, M.T. and Coene, W.M.J. (1993) *Ultramicroscopy* **48**, 77.

Parsons, J.R., Johnson, H.M., Hoelke, C.W., and Hosbons, R.R. (1973) *Phil. Mag.* **29**, 1359.

Rasmussen, D.R., Summerfelt, S.R., McKernan, S.R., and Carter, C.B. (1995) *J. Microscopy* **179**, 77.

Reimer, L. (1993) *Transmission Electron Microscopy; Physics of Image Formation and Microanalysis,* 3rd edition, Springer-Verlag, New York.

Rose, H. (1990) *Optik* **85**, 19.

Rose, H. (1991) in *High Resolution Electron Microscopy: Fundamentals and Applications* (Eds. J. Heydenreich and W. Neumann), p. 6, Institut für Festkörperphysik und Electronmikroskopie, Halle/Saale, Germany.

Shannon, C.E. and Weaver, W. (1964) *The Mathematical Theory of Communication,* University of Illinois Press, Urbana, Illinois.

Thon, F. (1975) in *Electron Microscopy in Materials Science* (Ed. U. Valdrè), p. 570, Academic Press, New York.

Van Dyck, D. (1992) in *Electron Microscopy in Materials Science,* p. 193, World Scientific, River Edge, New Jersey.

Van Dyck, D. and de Jong, A.F. (1992) *Ultramicroscopy* **47**, 266.

Van Landuyt, J., Van Tendeloo, G., and Amelinckx, S. (1991) in *High Resolution Electron Microscopy: Fundamentals and Applications* (Eds. J. Heydenreich and W. Neumann), p. 254, Institut für Festkörperphysik und Electronmikroskopie, Halle/Saale, Germany.

Image Simulation

29

29.1. Simulating Images 485
29.2. The Multislice Method 485
29.3. The Reciprocal-Space Approach 485
29.4. The FFT Approach 487
29.5. The Real-Space Approach 488
29.6. Bloch Waves and HRTEM Simulation 488
29.7. The Ewald Sphere Is Curved 489
29.8. Choosing the Thickness of the Slice 490
29.9. Beam Convergence 490
29.10. Modeling the Structure 491
29.11. Surface Grooves and Simulating Fresnel Contrast 492
29.12. Calculating Images of Defects 493
29.13. Simulating Quasicrystals 494
29.14. Bonding in Crystals 496

CHAPTER PREVIEW

When we need to obtain information about the specimen in two directions, we need to align the specimen so the beam is close to a low-index zone axis. If the HRTEM image information is going to be directly interpretable, the specimen must be oriented with the incident beam exactly aligned with both the TEM's optic axis and the specimen's zone axis. Thus we will have many reflections excited and the simple two-beam analysis of Chapter 27 cannot be used.

A method for modeling the contrast of images obtained under these conditions was developed by Cowley, Moodie, and their co-workers, principally at Melbourne and Arizona State University (ASU), in a series of classic papers beginning with that by Cowley and Moodie (1957). Fortunately, the growing interest in HRTEM has coincided with the availability of increasingly powerful computers which can handle the extensive calculations.

There are several software packages available commercially, so there is little reason for most users to re-invent the wheel. However, the different packages do not necessarily perform the calculations in the same way: one may be more appropriate for your application than others. Since these packages essentially operate as "black boxes," it is also reassuring to simulate images from the same structure using different packages (unless they don't give the same answers).

One point to keep in mind as you work through the literature is that this subject already has a lot of history. We will point out some of the things that have been done, and in some cases continue to be done, for historical reasons.

Image Simulation

29.1. SIMULATING IMAGES

The idea of simulating HRTEM images arose because of the realization that the loss of phase information when we form an experimental intensity map means that we can't go back from the image to the structure. Instead, we assume a structure (perfect crystal or crystalline material containing defects), simulate the image, see how closely the simulated image resembles the experimental image, modify the structure, and repeat the process. The only difficulty is that the image is sensitive to several factors:

- The precise alignment of the beam with respect to both the specimen and the optic axis.
- The thickness of the specimen (as we saw in Chapter 27).
- The defocus of the objective lens.
- Chromatic aberration, which becomes more important as t, the thickness, increases.
- Coherence of the beam.
- Other factors: one example would be the intrinsic vibration in the material, which we take account of through the Debye–Waller factor.

In principle, we could have the same image from two different structures. So obviously, this is the tricky part!

29.2. THE MULTISLICE METHOD

The basic multislice approach used in most of the simulation packages is to section the specimen into many slices, which are normal to the incident beam.

There are different methods for actually performing the multislice calculation. The different approaches have been developed for several reasons. Some try to optimize the use of available hardware. Others were written with the intention of providing a convenient method of simulating DPs using the same program. At least one package was written to make use of a popular personal computer with a user-friendly interface. The principal methods for performing these calculations are:

- The reciprocal-space formalism.
- The FFT formalism.
- The real-space approach.
- The Bloch-wave approach.

We'll go through the special features of each approach. The software packages which are readily available are listed in Section 1.5.

29.3. THE RECIPROCAL-SPACE APPROACH

We project each slice onto a plane somewhere in the slice (usually the top, bottom, or middle), giving a projected potential for that slice, and we call this the phase grating. We then calculate the amplitudes and phases for all the beams which will be generated by the incident beam interacting with the first projection plane. We could think of this as being a many-beam image calculation for a single slice. We then allow all these beams to propagate down the microscope in free space until they meet the next plane. The scattering calculation is now repeated for all the beams incident on this plane. This calculation produces a new set of beams which propagate through free space to the next plane, and so on. The process is summarized in Figure 29.1.

One point which you must remember: scattering by the phase grating does not just produce Bragg beams. It is

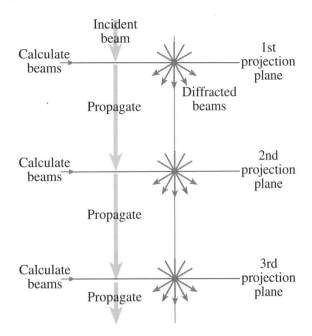

Figure 29.1. The potential within a slice is projected onto the first projection plane; this is the phase grating. We calculate the amplitudes and phases for all the beams generated by interacting with this plane and then propagate all the diffracted beams through free space to the next projection plane, and repeat the process.

crucial to keep track of the scattering in *all* directions. All of these beams will be incident on the next phase grating. So we don't just have Bragg beams, we *sample* all of reciprocal space.

A calculation based on a 128 × 128 array will impose a limit of ~4096 on the number of "beams" which can be included in the calculation. This number might appear large, especially when you form a [110] HRTEM image of Si with six Bragg beams (plus the 0 beam) but, particularly for imperfect crystals, this number will be inadequate.

Aside: Why do we need to consider regions of **k** space between the Bragg beams? In other words, why do we need to sample all of reciprocal space? The answer is that the Bragg beams contain information about the periodic structure, but all of the information from defects, i.e., nonperiodic structure, is contained *between* the Bragg spots, though it will generally be quite close to them.

Essentially, the multislice method considers three components:

- ψ   describes the *electron* wave.
- P   is the propagator of the electron wave in free space: the *microscope*.
- Q   is the phase grating: the *specimen*.

The process can be described by this equation:

$$\psi_{n+1}(\mathbf{k}) = [\psi_n(\mathbf{k}) P_{n+1}(\mathbf{k}) \otimes Q_{n+1}](\mathbf{k}) \quad [29.1]$$

where $\psi_{n+1}(\mathbf{k})$ is the wave function in reciprocal space at the exit of the $n+1$ slice and the symbol $\otimes$ denotes a convolution; $P_{n+1}(\mathbf{k})$ is the propagator for the $n+1$ slice. In other words, this is expressing the Fresnel diffraction phenomenon for this one slice because we are making a near-field calculation. (Look back to Chapter 2 for a discussion of near-field versus far-field.) Similarly, $Q_{n+1}(\mathbf{k})$ is the phase-grating function; it is a transmission function, for the $n+1$ slice.

The three functions $\psi(\mathbf{k})$, $P(\mathbf{k})$, and $Q(\mathbf{k})$ are all functions in reciprocal space, so this approach is referred to as the reciprocal-space formulation. Notice that the functions are all two-dimensional arrays. We can think of the different terms as being diffracted beams within the specimen. We can easily insert a circular objective aperture of radius **r**; we just require that all values of $\psi(\mathbf{k})$ are zero for $\mathbf{k} > \mathbf{k}_r$.

To give you an idea of the complexities involved, consider what values of $Q(\mathbf{k})$ you must use in the calculation. $Q(\mathbf{k})$ must go out twice as far as $\psi(\mathbf{k})$ or $P(\mathbf{k})$ in reciprocal space. You can understand why by considering Figure 29.2. If you represent the number of beams from slice $Q_{n-1}(\mathbf{k})$ as $F(\mathbf{k}')$, then $Q(\mathbf{k} - \mathbf{k}')$ must go out to $k = -4$ because, when you multiply these two functions to give $\psi(\mathbf{k})$, you can produce $k = -2$ by using $k = -4$ in Q and $k = +2$ in F as in Figure 29.2B. Putting this into an equation we have

$$\sum_{\mathbf{k}'} F(\mathbf{k}')Q(\mathbf{k} - \mathbf{k}') = \psi(\mathbf{k}) \quad [29.2]$$

where

$$F(\mathbf{k}) = \psi(\mathbf{k}) P(\mathbf{k}) \quad [29.3]$$

The function $Q(\mathbf{k})$ is a "probability map." What we are doing here is using the convolution to describe multiple scattering.

We can illustrate the complexity of the calculation by considering a 128 × 128 array for $Q(\mathbf{k})$ using SHRLI81 (see Section 1.5). The maximum value for (k_x, k_y) is only (31, 31), but even so, the number of diffracted beams is nearly 4096. Remember, we usually just use the seven inner beams in, e.g., the Si <110> DP, as we saw in Figure 27.3; most of the beams in our calculation are not Bragg beams. However, you will remember that the information concerning defects in crystals is contained in the regions between the Bragg spots in the diffraction pattern, so it does make sense. Specific examples of $Q(\mathbf{k})$, including nu-

29 ■ IMAGE SIMULATION

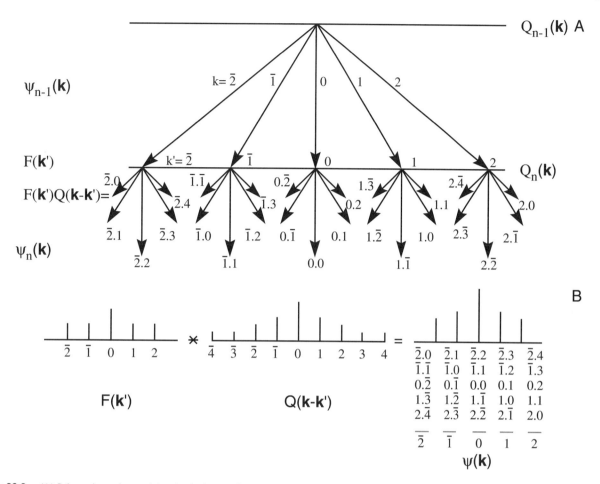

Figure 29.2. (A) Schematic used to explain why, in the one-dimensional case, $Q(\mathbf{k})$ must take account of twice as many $\mathbf{k}$ values as $\psi(\mathbf{k})$ or $P(\mathbf{k})$. Consider wave $k = \bar{2}$ from $Q_n(\mathbf{k})$: to produce wave $+2$ at this point you need to add 4 to $\bar{2}$ and similarly for every possible wave in slice $Q_n(\mathbf{k})$. As summarized in (B) $Q(\mathbf{k} - \mathbf{k}')$ extends from $\bar{4}$ to $+4$ so that $\psi(\mathbf{k})$, which we want, extends from $\bar{2}$ to $+2$, including all possible combinations of $\mathbf{k}'$ and $\mathbf{k}$.

merical computations of the phase change per slice, are given by Barry (1992)

29.4. THE FFT APPROACH

We can recast equation 29.3 to maximize the efficiency of the computer in using fast Fourier transform (FFT) routines. In this equation, F and F^{-1} tell us to take the Fourier transform or the inverse transform of the function inside the brackets

$$\psi_{n+1}(\mathbf{k}) = F\left\{F^{-1}[\psi_n(\mathbf{k})P_{n+1}(\mathbf{k})]q_{n+1}(x)\right\} \quad [29.4]$$

In this equation, $q_{n+1}(\mathbf{r})$ is the real space form of $Q_{n+1}(\mathbf{k})$, i.e., it is the inverse Fourier transform of $Q_{n+1}(\mathbf{k})$. So $q(\mathbf{r})$ is a real-space phase grating. Now we can look at some numbers for the calculation and take $Q(\mathbf{k})$ as a 128×128 array to keep the calculation small. The main steps carried out by the computer are:

- Multiply $\psi_n(\mathbf{k})$ by $P_{n+1}(\mathbf{k})$: that is a 64×64 array times another 64×64 array. Remember that you are limited to 64 points because the Q array must be twice as large in all directions in $\mathbf{k}$ space.
- Take the inverse Fourier transform of the result.
- Multiply this new result by $q_{n+1}(\mathbf{r})$, which is the 128×128 array.
- Fourier transform the final result and set all values outside the inner 64×64 array equal to zero so that you can repeat the process for the next slice.

You will notice that this example used a square array. In modern programs, we are not restricted even to using powers of 2 but this helped the original FFT routines. You will see the value of this advance when we examine some defect calculations later. If you are interested in the mechanics of the FFT routine and other aspects of this simulation approach, the article by O'Keefe and Kilaas (1988) is required reading.

29.5. THE REAL-SPACE APPROACH

As we noted earlier, image simulation used to be limited by your budget, i.e., by your computer. The real-space approach was developed, in part, to decrease the time needed for the calculations by using our knowledge that $P(\mathbf{r})$ is strongly peaked in the forward direction. In our notation, the Coene and Van Dyck (1984a,b) method for calculating $\psi(\mathbf{x})$ can be expressed by the equation

$$\psi_{n+1}(\mathbf{r}) = [\psi_n(\mathbf{r}) \otimes P_{n+1}(\mathbf{r})] q_{n+1}(\mathbf{r}) \quad [29.5]$$

where $P_{n+1}(\mathbf{r})$ is now the propagator in real space and $q_{n+1}(\mathbf{r})$ is again the real-space phase grating. Once you have written this, it's all computing, which is a substantial task since the size of the multislice calculation is the size of the largest array, i.e., $Q(\mathbf{k})$ or $q(\mathbf{x})$.

29.6. BLOCH WAVES AND HRTEM SIMULATION

Although we saw in Chapters 14 and 15 that electrons propagate through crystalline specimens as Bloch waves, the multislice method we've described so far is essentially a "diffracted-beam" approach. In two classic papers Fujimoto (1978) and Kambe (1982) showed that, for the perfect crystal, the HRTEM may be understood simply in terms of images of Bloch waves. The key point is that, although a large number of diffracted waves are formed, only a small number of Bloch waves determine the appearance of the image, providing the crystal has a sufficiently high symmetry. Following Kambe's "simple" example we consider the case where only three Bloch waves i, j, and k are significant. Let's assume that Bloch waves i and j are in phase at a thickness $z = D$. Then we have

$$e^{ik_z^{(i)}z} = e^{ik_z^{(j)}D} \quad [29.6]$$

(Don't confuse the kth Bloch wave with the **k**-vector!)

Using our expression for ψ, namely

$$\psi(\mathbf{r}) = \sum_i C^{(i)} \phi^{(i)}(x, y) e^{ik_z^{(i)}z} \quad [29.7]$$

and the normalization rule

$$\sum_i C^{(i)} \phi^{(i)}(x, y) = 1 \quad [29.8]$$

we can therefore express ψ at $z = D$ in terms of our three Bloch waves

$$\psi(x, y, D) = \left[C^{(i)} \phi^{(i)} + C^{(j)} \phi^{(j)} \right] e^{ik_z^{(i)}D} + C^{(k)} \phi^{(k)} e^{ik_z^{(k)}D} \quad [29.9]$$

We rearrange this equation so that we can extract the phase factor $e^{ik_z^{(i)}z}$ ($= e^{ik_z^{(j)}D}$). We write

$$\psi(x, y, D) = \left[1 - C^{(k)} \phi^{(k)} \right] e^{ik_z^{(i)}D} \\ + C^{(k)} \phi^{(k)} e^{i\left(k_z^{(k)} - k_z^{(i)}\right)D} e^{ik_z^{(i)}D} \quad [29.10]$$

$$\psi(x, y, D) = e^{ik_z^{(i)}D} \left[1 + \beta_{ik}(D) C^{(k)} \phi^{(k)} \right] \quad [29.11]$$

where we've defined a new parameter β given by

$$\beta_{ik}(D) = e^{i\left(k_z^{(k)} - k_z^{(i)}\right)D} - 1 \quad [29.12]$$

These equations tell us that if any two of the Bloch waves (here they are i and j) are in phase, then the amplitude of the wave at the exit surface is determined by the third Bloch wave.

If the third Bloch wave is also nearly in phase, we have a relation like equation 29.6 but with i, j, and k all equal. Then we can approximate $\beta_{ik}(D)$ by

$$\beta_{ik}(D) \simeq i\left[\left(k_z^{(k)} - k_z^{(i)}\right)D + 2n\pi\right] = i\gamma_{ik}(D) \quad [29.13]$$

Now we've defined another factor γ_{ik}. If you plug this expression back into equation 29.11, you see we have a pure phase object. All the diffracted beams will be shifted in phase by $\pi/2$.

Now you can test the effects of how we change k. Consider what conditions this will really correspond to using equations 29.11 and 29.13.

- If k is such that the phase of Bloch wave k is ahead of i and j (which were equal), then you'll see a "negative" image of $C^{(k)}\phi^{(k)}$. A "delayed" k gives us the "positive" image.
- For the Ge <110> zone axis, HRTEM image at 100 kV, only three Bloch waves are strongly excited.

The relationship to the Bloch-wave contours in Chapter 14 is clear. Using this information and the projected potential shown in Figure 29.3, Kambe calculated the Bloch-wave amplitudes and the two ideal images of the Bloch waves: one positive and the other negative. In the calculation of different images for increasing thickness, several images corresponding to a single Bloch wave can be predicted and identified, as shown in the figure. At other thicknesses the images form by a combination of Bloch waves. So, what can we learn?

- For a perfect crystal, you may need as few as three Bloch waves to give the essential features of an HRTEM zone-axis image.
- There is a direct connection between the WPOA and the propagation of Bloch waves.

We saw in Chapter 14 that the electron propagates as Bloch waves inside the crystal. The multislice approach, which we usually use to simulate HRTEM images, is actually a very elegant form of brute force. The reason we don't use Bloch waves is that our specimens are not perfect. However, EMS (see Section 1.5) does give you the option of using this approach.

29.7. THE EWALD SPHERE IS CURVED

When you are using the TEM, some other complications arise because the Ewald sphere is curved:

- If you align the beam exactly parallel to a zone axis, **s** will be nonzero for every Bragg reflection. In fact, it will also be different for each type of reflection.

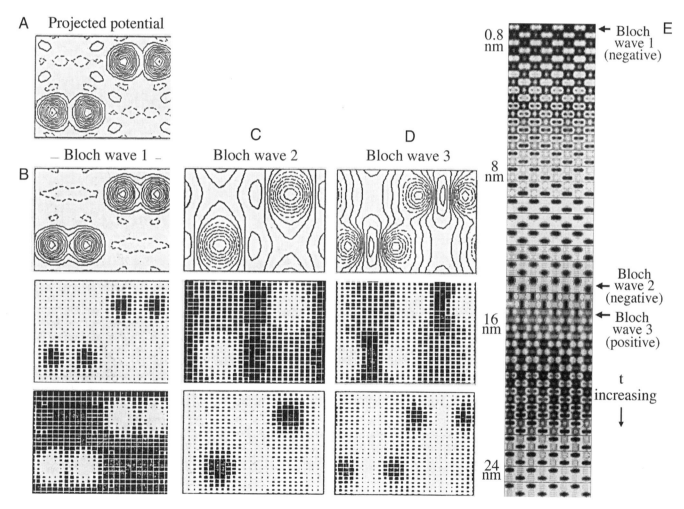

Figure 29.3. (A) The projected potential for Ge where the contour lines represent changes in potential of -10 eV, and the dashed lines are positive values; (B) the amplitudes for Bloch waves 1, 2, and 3 for 100-keV electrons; (C) ideal positive images of the Bloch waves; (D) ideal negative images of the Bloch waves; (E) the thickness dependence of the lattice image.

- If you do not align the beam exactly parallel to the zone axis, then **s** will also be slightly different for each reflection in that zone.
- If you change the wavelength of the electrons, the radius of the sphere changes.
- If you converge the beam, then you'll add a thickness to the Ewald sphere.

The point is, knowing precisely what the correct values are to put in the program will also require thought and work.

29.8. CHOOSING THE THICKNESS OF THE SLICE

So far, we've just cut the specimen into slices in the computer without considering how thick each slice should be, or even whether they should all be the same. If all the slices are the same, then there can be no information about the z-direction. Although HOLZ lines are not important for the simulation of HRTEM images, some of the programs we are discussing can now just as readily be used to simulate CBED patterns and HOLZ lines. So, following the philosophy of attacking problems with different techniques, you should be aware of these limitations, since it is easy to overlook the simplifications you made once you see the computed image. You should remember that when you are studying a material with a large unit cell, the reciprocal lattice spacing will be short in the beam direction, so HOLZ effects come into play sooner.

Consider the different methods for making the slice:

- You could calculate the projected potential for a thick slice and then do n calculations with slices which are $1/n$ times this thickness.
- A better approach would be to subdivide the cell into layers of atoms, create a different grating for each of these layers, and then run the program with the sequence.

For example, if the beam is aligned along the [111] direction of an fcc crystal, then you would have three identical gratings displaced relative to one another, giving the ABC stacking of close-packed planes. This approach would allow you to test for the effect of a real error in the stacking sequence normal to the beam. Even this point can be a bit difficult. In general, you orient the beam to be parallel to a particular zone axis [UVW] so that the planes in that zone are parallel to the beam (so our projection works). If the material is not cubic, you will not generally have a low-index plane normal to the beam to make this slice.

29.9. BEAM CONVERGENCE

When you are recording HRTEM images, you need to keep exposure times short. So, if you don't use parallel illumination, you have to take account of the beam convergence when simulating the images. O'Keefe and Kilaas (1988) (see also Self and O'Keefe 1988) have developed one approach to address this problem. If the beam actually has some convergence, then the diffraction spots will be disks, as illustrated in Figure 29.4, so you need to simulate disks in the DP. Experimentally, the large objective aperture admits many disks, so in the simulation routine you should sample each disk at many points. This means the program needs to calculate the image at each of these convergence angles and average all the resulting images. Of course, the objective aperture is easily applied in the computer. If you choose 49 points, you can make the sampling interval in reciprocal space ≤ 0.1 nm^{-1}. It is instructive to examine just how much work is necessary to sample the 49 points.

We can start by writing the usual expression for χ, the phase change due to the objective lens

$$\chi = \pi \Delta f \lambda u^2 + \pi C_s \lambda^3 \left(\frac{u^4}{2}\right) \qquad [29.14]$$

Then differentiate this with respect to the variable u

$$\frac{d\chi}{du} = 2\pi \left(\lambda u \Delta f + C_s \lambda^3 u^3\right) \qquad [29.15]$$

This equation tells us that if u changes by δu, then χ changes by

$$\delta \chi = 2\pi \lambda \left(u \Delta f + C_s \lambda^2 u^3\right) \delta u \qquad [29.16]$$

Now we choose $\delta \chi$ so that

$$\delta \chi < \frac{2\pi}{n} \qquad [29.17]$$

where n will allow us to determine the maximum change in χ between two points in the disk. For example, if $n = 12$, then the maximum value of $\delta \chi$ is 30°. Combining equations 29.15 and 29.17, we can write

$$\delta u = \left[n \lambda u \left(\Delta f + C_s \lambda^2 u^2\right)\right]^{-1} \qquad [29.18]$$

If we plot χ versus u (or play with equation 29.15 and its derivative), then we find a minimum at

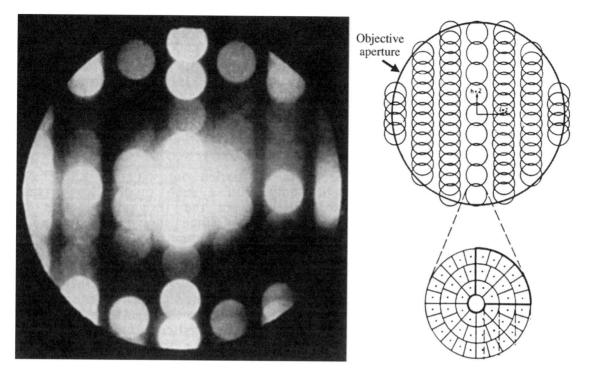

Figure 29.4. Disks in the DP from a crystal of $Nb_{12}O_{29}$. The computer simulation can divide each disk into many sectors and simulate the image for each sector, as shown in the schematic, excluding sectors which are intersected by the objective aperture.

$$\Delta f = -C_s \lambda^2 u^2 \quad [29.19]$$

and an inflection at

$$\Delta f = -3C_s \lambda^2 u^2 \quad [29.20]$$

So the simulation program can check to find the smallest δu at an inflection point, which equations 29.18 and 29.20 tell us is

$$\delta u = -\left[\frac{27C_s}{(\Delta f)^3}\right]^{\frac{1}{2}} \left(\frac{1}{2n}\right) \quad [29.21]$$

The value of δu therefore depends on both C_s and Δf.

> Remember that all this calculation takes place in that black box!

You can also appreciate the relevance of this type of approach if your disks actually intersect the objective aperture, as shown in Figure 29.4. Put another way, you can learn two lessons from this analysis:

- Always try to minimize the convergence of the beam when recording HRTEM images.

- Use an aperture which does *not* cut through the diffraction disks.

29.10. MODELING THE STRUCTURE

To simulate any HRTEM image, you need a unit cell. If you are only concerned with perfect crystals, then your program should have all the space groups already included so that you only need to add the lattice parameters (lengths and angles) and the occupied sites for your material. If you are interested in simulating images from defects, then you have to create a new unit cell which must be sufficiently large that it will not add effects due to the edges. There are many ways to create this defect unit cell. You can input from other programs, such as those performing atomistic modeling of defects, or create your own starting structure. In either case, you will need to move atoms, either manually or following a rule you've selected for image matching, to optimize the match between your experimental series of through-focus images and the simulated images.

At some stage, you will find it useful to combine different slices, as when simulating grain boundaries with or without a surface groove, or modeling large complex

29.11. SURFACE GROOVES AND SIMULATING FRESNEL CONTRAST

The analysis of interfaces by the Fresnel-fringe technique, which we introduced in Chapter 27, illustrates the importance of image simulation and emphasizes that it is not just for HRTEM. The calculation is complicated for several reasons, as shown in Figure 29.5A:

- The potential change at the interface is probably not abrupt.
- The potential depends on the detailed structure of the interface.
- During preparation, TEM specimens may be preferentially damaged at grain boundaries, giving rise to surface grooves.

If you use a thicker specimen you'll reduce the effect of surface grooves on any Fresnel fringes, but in practice your foil thickness is usually limited (~ 20 nm), since you need to view the boundary exactly edge on. Even for foils this thick, surface grooves can influence the projected potential considerably. If we assume that the bulk has a mean inner potential $V = 20$ V, and take a typical potential drop for an intergranular film to be ~1 V, then the total projected potential drop for a 20-nm-thick foil would be the same as that caused by a pair of grooves at the top and bottom surfaces, which are only 0.5 nm deep. Although the surface groove may be partly filled with a second phase, the effect on the Fresnel fringes can still be substantial.

We can examine Fresnel fringes using different methods. In all of them, we describe the potential at the interface in terms of the projected potential drop $\Delta V_p = t\Delta V$, an inner width a, an outer width a_0, and a "diffuseness," δ, defined by

$$a_0 = (1 + \delta)a \qquad [29.22]$$

These parameters are shown in Figure 29.5B. Then we construct models of a foil with a surface groove at the edge-on interface by combining such potentials.

The models: Values of $\delta = 0.5$ and $\delta = 0.2$ represent shallow and steep surface grooves respectively. The total projected potential drop can be due to a real change in V or a change in t. A groove without a film implies

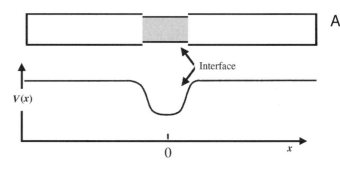

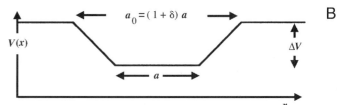

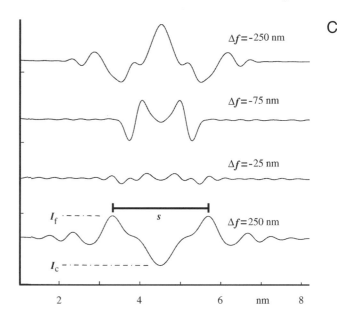

Figure 29.5. (A) Schematic of a grain boundary containing a layer of material with a different inner potential; (B) one model used to represent such a grain boundary giving variable parameters a, a_0, and δ; (C) a typical set of simulated Fresnel-fringe intensity profiles at increasing Δf: s is the distance between the first two fringes, I_c and I_f are the intensities of the central and first fringe, respectively.

$a = 0$. If $a = 1$ nm and $a_0 = 1.5$ nm, the model could correspond to two different situations:

- If the atoms at the interface relax, then the atomic density at the interface will usually be reduced. This occurs at both structured interfaces and those where a layer of glass is present.
- The surface grooves at the interface.

What image simulation shows is that the relative shapes and sizes of these models are more important than the actual dimensions. Therefore, we can give most of the following analysis in terms of dimensionless quantities. Inner potentials are typically 5–10 eV. Except for very small defocus values, i.e., $\Delta f \sim t$, we find that the distribution of the potential through the foil is not important. Usually, the projected potential at the interface is lower than that in the bulk. However, the opposite situation can occur, e.g., when a Bi_2O_3-rich phase is present at interfaces in ZnO. When we discuss the calculated profiles, the term "interface width" will be used for the parameters a and a_0, whether they actually correspond to an intergranular film, a surface groove, or otherwise.

> In the case of Fresnel fringes from an edge, the distance from the edge to the first fringe is proportional to $(\lambda \Delta f)^{1/2}$. The fringe spacing, s_f, can be extrapolated to zero defocus to obtain the interface width, based on the relation $(s_f - a) \propto \sqrt{(\lambda \Delta f)}$.

This relation (Clarke 1979) only holds when a is large and Δf is relatively small; then the fringes from each "edge" at the interface are independent. We observe the minimum fringe spacing at small values of defocus and this spacing can be used to provide a measure of the interface width. For more details on the simulation of Fresnel fringes, we refer you to the original articles and the papers by Taftø et al. (1986), Rasmussen and Carter (1990), Stobbs and Ross (1991), and Zhu et al. (1995).

In practice, the analysis of Fresnel fringes is impaired not only by specimen artifacts such as surface grooves, but also by various sources of noise, which all add to the uncertainty of measurements, especially at low defocus. For diffuse interfaces, the contrast decreases rapidly as Δf approaches zero (Figure 29.5C), and measurements of the fringe spacing are increasingly susceptible to noise and artifacts. You can always use larger defocus values and thus obtain higher contrast. However, without prior knowledge about the shape of the potential drop, e.g., the diffuseness, you can't reliably determine the interface width by measuring the fringe spacing alone. Since the fringe spacing is dominated by the outer width, a_0, you may easily overestimate the interface width. The atomic density in a region close to the boundary is often reduced, even if the boundary is structured, so you can easily misinterpret the image as showing the presence of an intergranular film when it is actually film-free (Simpson et al. 1986).

The region of defocus, where the central fringe shows little contrast, provides complementary information to the fringe spacings, so it is more sensitive to the inner width.

> The conclusion is that you must use all the information in the image to characterize the shape of the potential well.

From this discussion, you'll appreciate that, before you can completely understand the effect of any intergranular films, you must estimate the extent to which surface grooves are present in your specimen. Shadowing (e.g., using platinum or gold) may provide evidence for surface grooves, but in the case where the surface groove is already filled (e.g., if your specimen was coated with carbon), this technique won't work.

To summarize, this discussion gives us a method for analyzing Fresnel fringes from a grain boundary. We can draw some conclusions:

- To interpret the contrast from Fresnel fringes at grain boundaries, you must simulate images of many different interface models. In particular, it is essential that you consider the possibility of artifacts such as surface grooving. Even a rather "flat" or diffuse surface groove may influence the fringes in some range of defocus values.
- Both the fringe spacing and the central fringe intensity depend on the shape of the potential well and are sensitive to surface grooving.
- The interface width, which you can infer from the fringe spacing, is dominated by the outer width of a diffuse interface.
- A direct match with the s_f–a curve (or with similar simulated curves when the assumptions employed here fail) leads to a better estimate of the average interface width, but cannot give you much information on the shape of the potential well.
- Determining when the central fringe is weak (the range Δf) gives complementary information on the interface width which, in combination with the estimate based on the fringe spacing, you can use to evaluate the diffuseness of the potential well.

29.12. CALCULATING IMAGES OF DEFECTS

When we simulate HRTEM images of perfect crystals, we only need input the unit cell and the program generates the rest of the specimen. If we want to calculate the image of a defect, we have to use the same approach: we set up a unit cell to contain the defect and the program treats it like any other unit cell. This is known as the *periodic continuation method* for defect calculation. What we've actually done is shown in Figure 29.6: there is an array of defects throughout our specimen in all directions. We need to know two things:

- To what extent does this ordered array introduce artifacts in the image?
- Have we created interfaces where the "cells" join which may influence our image?

An example of a supercell for a grain boundary is shown in Figure 29.6. This figure illustrates clearly how we can create a cell which is more suitable for this periodic continuation by including two defects in a single supercell. As shown in this figure, the periodic continuation then not only creates many other grain boundaries but also makes them very long. If we don't match the crystals exactly at the edges of the supercell we create a different "ghost" boundary.

You can see that this really can be a problem by considering the DP which our new cell would produce. We are calculating the image of a small part of a periodic array of interfaces. Periodic arrays in real space produce rows of extra spots in reciprocal space. If we include these spots in forming the image, we should change the image. The solution for image simulation is quite simple; make the supercell wider and wider until the change in the image detail is

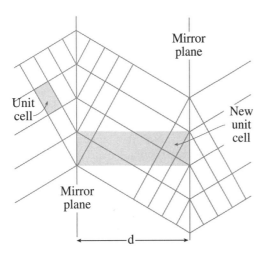

Figure 29.6. The periodic continuation technique illustrating how an artificial unit cell can be constructed to contain two grain boundaries, thus allowing the HRTEM image to be simulated. The distance (d) between the two interfaces can be varied to check for overlap artifacts.

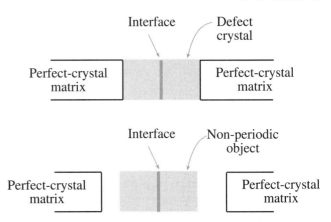

Figure 29.7. In the real-space patching method, the defect crystal (in this case the interface and several adjacent layers) is a nonperiodic object which is surrounded by a perfect-crystal matrix.

less than some specified limit. However, don't try to interpret the data in the calculated DP without consulting the paper by Wilson and Spargo (1982).

An alternative approach to the periodic continuation approach has been developed by Coene *et al.* (1985) and is called the real-space patching method. This method uses the "real-space" image simulation approach to perform the calculation. The structure you want to simulate can be divided into a number of different "patches" as illustrated in Figure 29.7; the image from each patch is calculated for a slice and then the patches are joined together. The key, of course, is that you must correctly take account of what happens at the edge of each patch. This means that each patch needs some information about the neighboring patches. Assuming (correctly) that this can be done, you can appreciate the nice feature of this approach: we avoid the artificial interference effects due to the array of defects that would be produced by the periodic continuation technique. As explained by Coene *et al.* (1985), the defect does not now "see" its own image, it only sees the perfect matrix on all sides.

29.13. SIMULATING QUASICRYSTALS

There are several problems in simulating HRTEM images of quasicrystals, not least of which concerns which model you should use. Several models have been reviewed by Shoemaker (1993) and the possibilities are illustrated by the work of Beeli and Horiuchi (1994), who used a combination of 10 layers in the multislice calculation. The layers are made up from the planar structures shown in Figure 29.8. The final structure (shown in Figure 29.8A) is made up of two sets of five layers. The first set of layers

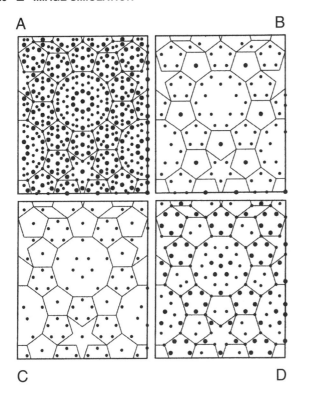

Figure 29.8. Projections used to simulate images of $Al_{70}Mn_{17}Pd_{13}$ quasicrystals. (A) Combination of all the layers; the (B, C, D) layers are used to contribute (repeatedly) to (A). The edges of all the tiles are 0.482 nm. The large circles denote Al atoms.

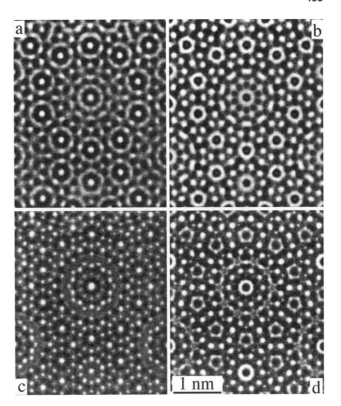

Figure 29.9. Four simulated images of the model constructed from the layers shown in Figure 29.8 using only Al and Mn atoms. The thickness is 3.77 nm, which corresponds to three periods in the beam direction. The values of Δf are (a) 0 nm, (b) 46 nm, (c) 88 nm, and (d) 124 nm.

is B-C-D-C-B in this figure. The second set of five layers is constructed from the first by using the screw symmetry of the structure; the screw axis has 10_5 screw symmetry. The supercell used was 3.882 nm by 3.303 nm, which was chosen to contain a complete decagonal cluster that is 2.04 nm in diameter and the center part of a pentagon tile. As you realize, the problems in such an image calculation are increased because the quasicrystal does not have translational symmetry, but you must impose such a symmetry to do the calculation. The calculation was then carried out for thicknesses up to 10 nm.

The results of such calculations with only Al and Mn atoms are illustrated in Figure 29.9. The edges of the cells are essentially artificial because, as we just noted, the structure used in the calculation is a "unit cell." In spite of these difficulties Beeli and Horiuchi could conclude that the image match was much improved when Pd atoms were included to replace some of the Mn atoms in the D layer and Al atoms in the B-C layers, with the results shown in Figure 29.10.

Another illustration of the success of HRTEM comes from the work of Jiang *et al.* (1995) on quasicrystals with eightfold symmetry. Here the multislice calculation could again be made using a relatively simple sequence of four layers ABAB′, where the layers are at $z = 0$, 0.25, 0.5, and 0.75. The structures of the A and B layers are shown in Figure 29.11 with the B′ layer being a 45°-rotated B layer, i.e., the B and B′ layers are again related by a screw axis, but this time it's an 8_4 screw axis.

- In each of these examples, it is possible to view the same structure parallel to an orthogonal axis.
- Quasicrystals do not have translational symmetry, but we pretend they do for thickness calculations and for the periodic continuation of the unit cell.

Our reason for showing so much detail on these rather esoteric materials is that they show what can be done using image simulation. Furthermore, they emphasize the important fact that, although we can construct the crystal using different layers and different sequences of layers, we always use a projection of the structure, to compare with the experimental image.

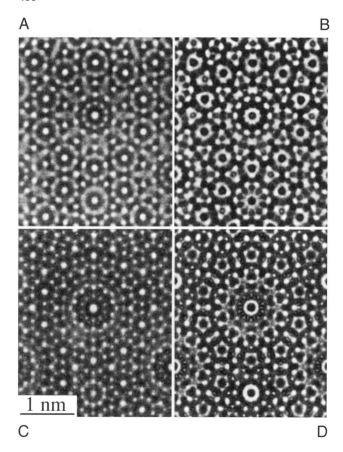

Figure 29.10. Examples of simulated images of the quasicrystal shown in Figure 29.8 but substituting Pd atoms for Mn atoms. The values of Δf are (A) 0 nm, (B) 48 nm, (C) 88 nm, and (D) 128 nm.

29.14. BONDING IN CRYSTALS

We mentioned early on that one problem we have with image simulation concerns the fact that atoms are bonded in different ways in different materials. The standard approach has been to use values for structure factors tabulated by Doyle and Turner (1968) and Doyle and Cowley (1974). These values were calculated using a relativistic Hartree–Fock (RHF) model for the atomic potential. An alternative approach is to relate the scattering factor for electrons (f_e) to that for X-rays (f_x) using the Mott equation, or to use a more sophisticated atomic potential known as the relativistic Hartree–Fock–Slater (RHFS) model. Carlson *et al.* (1970) give tabulated results while Tang and Dorignac (1994) have made detailed comparisons for HRTEM imaging.

O'Keefe and Spence (1994) have re-examined the meaning of the mean inner potential. One of the reasons that you need to understand this concept is that we often link data from X-ray diffraction and data from electron dif-

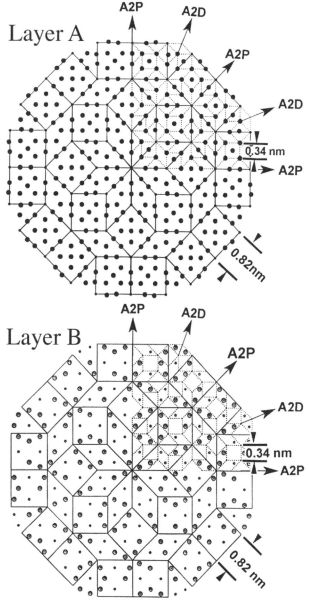

Figure 29.11. The model used to simulate quasicrystals with eightfold symmetry. The structure for the simulation was constructed as a four-layer sequence ABAB' where the B and B' layers are related by an 8_4 screw axis.

fraction. As usual, computers are making it possible to do more elaborate calculations using other potentials.

While this is an evolving study, some important results have been obtained:

- The inner potential is very sensitive to bonding effects. O'Keefe and Spence discuss this result for MgO (large ionic component), Si (covalent), and Al (metallic).

- We are still not able to take full account of bonding effects, which could be important for HRTEM images.

This paper by O'Keefe and Spence is highly recommended reading for those who have a strong physics background but think TEM is a "known" subject!

CHAPTER SUMMARY

If you are going to do HRTEM imaging, you must be prepared to use image simulation to assist you in interpreting your images. If you want to do quantitative imaging, simulation is an essential component of the process, as we'll see in Chapter 30. Most materials scientists using TEM will want to use one of the established software packages we listed in Section 1.5. There are several important conclusions contained in this chapter:

- Make sure that you know all you can about your specimen. We illustrated the dangers with our discussion of grooved grain boundaries. You can waste too much time looking at artifacts caused by specimen preparation.
- Make sure that you know all you can about your TEM. You now have some idea of how many parameters are required by the simulation routines. Beware of the parameters which you did not measure for your machine. The program will need to use some value.
- Make sure that you accurately align your TEM before you record any images.
- If possible, use more than one program to simulate the images.
- Record a through-focus series and check for changes in Δf by repeating the first image.
- The fact that the thickness of your specimen varies can be a great asset provided you can determine that thickness, i.e., it gives you another variable.

The traditional method of using simulated images has often involved looking at a series of simulated images for different values of Δf and t and finding the best match with your experimental image. Clearly, this is not the ideal approach. Remember that the interpretation of HRTEM images may not be straightforward or unique. We must next compare the simulated images with those generated experimentally. This is the subject of the next chapter and is the basis of quantitative HRTEM.

REFERENCES

General References

Buseck, P.R., Cowley, J.M., and Eyring, L., Eds. (1988) *High-Resolution Transmission Electron Microscopy and Associated Techniques*, Oxford University Press, New York. An excellent resource covering more than HRTEM.

Horiuchi, S. (1994) *Fundamentals of High-Resolution Electron Microscopy*, North-Holland, New York. An ideal complement to Spence's book.

Kihlborg, L., Ed. (1979) *Direct Imaging of Atoms in Crystals and Molecules*, Nobel Symposium 47, The Royal Swedish Academy of Sciences, Stockholm. A classic collection of papers on HRTEM.

Krakow, W. and O'Keefe, M., Eds. (1989) *Computer Simulation of Electron Microscope Diffraction and Images*, TMS, Warrendale, Pennsylvania. A collection of defocused review articles.

O'Keefe, M.A. and Kilaas, R. (1988) in *Advances in High-Resolution Image Simulation in Image and Signal Processing in Electron Microscopy*, Scanning Microscopy Supplement 2. (Eds. P.W. Hawkes, F.P. Ottensmeyer, W.O. Saxton, and A. Rosenfeld), p. 225. SEM Inc., AMF O'Hare, Illinois. This article is another excellent introduction to the subject.

Spence, J.C.H. (1988) *Experimental High-Resolution Electron Microscopy*, 2nd edition, Oxford University Press, New York. This is *the* text for users of the HRTEM.

Specific References

Barry, J. (1992) in *Electron Diffraction Techniques* **1**, (Ed. J.M. Cowley), p. 170, I.U.Cr., Oxford Science Publication, New York.

Beeli, C. and Horiuchi, S. (1994) *Phil. Mag.* **B70**, 215.

Carlson, T.A., Lu, T.T., Tucker, T.C., Nestor, C.W., and Malik, F.B. (1970) Report ORNL-4614, ORNL, Oak Ridge, Tennessee.

Clarke, D.R. (1979) *Ultramicroscopy* **4**, 33.

Coene, W. and Van Dyck, D. (1984) *Ultramicroscopy* **15**, 41.

Coene, W. and Van Dyck, D. (1984) *Ultramicroscopy* **15**, 29.

Coene, W., Van Dyck, D., Van Tendeloo, G., and Van Landuyt, J. (1985) *Phil. Mag.* **52**, 127.

Cowley, J.M. and Moodie, A.F. (1957) *Acta Cryst.* **10**, 609.

Doyle, P.A. and Cowley, J.M. (1974) in *International Tables for X-ray Crystallography*, Vol. **IV**, pp. 152–173, Kluwer Academic Publ., Dordrecht, the Netherlands.

Doyle, P.A. and Turner, P.S. (1968) *Acta. Cryst.* **A24**, 309.

Fujimoto, F. (1978) *Phys. stat. sol. (a)* **45**, 99.

Jiang, J.-C., Hovmöller, S., and Zou, X.-D. (1995) *Phil. Mag.* **71**, 123.

Kambe, K. (1982) *Ultramicroscopy* **10**, 223.

Kouh, Y.M., Carter, C.B., Morrissey, K.J., Angelini, P., and Bentley, J. (1986) *J. Mat. Sci.* **21**, 2689.

O'Keefe, M. and Spence, J.C.H. (1994) Acta. Cryst. **A50**, 33.

Rasmussen, D.R. and Carter, C.B. (1990) *Ultramicroscopy,* **32**, 337.

Ross, F.M. and Stobbs, W.M. (1991) *Phil. Mag.* **A63**, 1.

Self, P.G. and O'Keefe, M.A. (1988) in *High-Resolution Transmission Electron Microscopy and Associated Techniques,* (Eds. P.R. Buseck, J.M. Cowley, and L. Eyring), p. 244, Oxford University Press, New York.

Shoemaker, C.B. (1993) *Phil. Mag.* **B67**, 869.

Simpson, Y.K., Carter, C.B., Morrissey, K.J., Angelini, P., and Bentley, J. (1986) *J. Mater. Sci.* **21**, 2689.

Taftø, J., Jones, R.H., Heald, S.M. (1986) *J. Appl. Phys.* **60**, 4316.

Tang, D. and Dorignac, D. (1994) *Acta. Cryst.* **A50**, 45.

Wilson, A.R. and Spargo, A.E.C. (1982) *Phil. Mag.* **A46**, 435.

Zhu, Y., Taftø, J., Lewis, L.H., and Welch, D.O. (1995) *Phil Mag. Lett.* **71**, 297.

Quantifying and Processing HRTEM Images

30

30.1. What Is Image Processing? 501
30.2. Processing and Quantifying Images 501
30.3. A Cautionary Note 502
30.4. Image Input 502
30.5. Processing Techniques 502
    30.5.A. Fourier Filtering and Reconstruction 502
    30.5.B. Analyzing Diffractograms 504
    30.5.C. Averaging Images and Other Techniques 506
    30.5.D. Kernels 507
30.6. Applications 507
    30.6.A. Beam-Sensitive Materials 507
    30.6.B. Periodic Images 508
    30.6.C. Correcting Drift 508
    30.6.D. Reconstructing the Phase 508
    30.6.E. Diffraction Patterns 508
    30.6.F. Tilted-Beam Series 510
30.7. Automated Alignment 511
30.8. Quantitative Methods of Image Analysis 512
30.9. Pattern Recognition in HRTEM 513
30.10. Parameterizing the Image Using QUANTITEM 514
    30.10.A. The Example of a Specimen with Uniform Composition 514
    30.10.B. Calibrating the Path of R 516
    30.10.C. Noise Analysis 516
30.11. Quantitative Chemical Lattice Imaging 518
30.12. Methods of Measuring Fit 518
30.13. Quantitative Comparison of Simulated and Experimental HRTEM Images 521
30.14. A Fourier Technique for Quantitative Analysis 523
30.15. Real or Reciprocal Space? 523
30.16. The Optical Bench 524

CHAPTER PREVIEW

In this chapter we will equate our title with the use of the computer. We will simply use image processing to extract more information from a TEM image than we can obtain by eye. In the past the optical bench was also used for this purpose, but the number of optical benches is negligible compared to the number of computers now found in every TEM lab. Optical benches did allow us to form DPs which we could then modify to produce a processed image. This analog approach has now largely been replaced by its digital counterpart. The computer can be much cheaper than the optical bench and is far more flexible. The number of software packages which are designed for, or can easily be adapted to, TEM is also growing.

We can use image processing to produce a clearer view of the image, for example by subtracting unwanted background detail, correcting for noise or drift, or removing artifacts. The big warning, though, is that, when removing one artifact, you must be very careful not to introduce others.

Although it's nice to see information more clearly, the unique feature of the computer approach is that we can *quantify* the data in any image and then normalize these data. Now we can directly compare the quantified experimental image with computer-simulated images. Although throughout this chapter we will be concerned with HRTEM images, most of what we say can be transferred directly to the analysis of diffraction-contrast images.

The other general point is that the ideas we'll discuss are also applicable to images derived from different sources. Once the data are in the computer, i.e., in digital form, the source becomes unimportant as far as processing possibilities are concerned. Examples of "images" which might be obtained from the TEM include X-ray or EELS maps, STEM images, TEM images, and CBED or BSE patterns.

Most of our discussion will concern the use of computers. All you need to know is how best to get the data into the computer, how to process it, what to do with the data, and how to display the result!

Quantifying and Processing HRTEM Images

30

30.1. WHAT IS IMAGE PROCESSING?

Image processing is essentially manipulating images. The topic arises in many fields, so we need to understand the words/jargon; we'll discuss the language of image processing as it is applied to TEM.

> The basic idea of image processing is changing images into numbers and manipulating the numbers.

Image processing is not only becoming more common, but is also finding new applications in many fields. Faster, more powerful computers and increased memory storage are making tasks possible which could not previously have been considered. As a result of this increasing user base, there are now many software packages available which can be used in microscopy; we listed some back in Section 1.5. These range from programs used widely in desktop publishing to those which have been custom designed for EM. The goals of image processing include that of quantitative microscopy. You must choose between the different packages, commercial and freeware, and match them to the computer available in your lab. One point to remember is that some very simple optical methods which don't rely on a computer can often be very helpful. The other point is that the eye is hard to beat.

There are many specialized books on this topic for the beginner or the expert; a selection is given in the references. The purpose of this chapter is to give a generalized overview. One problem in discussing this topic is that it is a very rapidly changing field. We will try to avoid specifics concerning particular programs but will mention these programs at the end of this chapter.

30.2. PROCESSING AND QUANTIFYING IMAGES

We process images primarily for two reasons:

- We may want to improve the appearance of an image, make it look sharper, more even in contrast, higher contrast, etc.
- We may want to quantify the information contained in the image.

Processing for improving the appearance of TEM images has been practiced for many years using such photographic techniques as "dodging," using "filters," selecting different emulsions, or varying the developer, etc. It is only recently that relatively powerful personal computers have become widely available, but the term "image processing" almost automatically implies the use of computers. Computer image processing will be the emphasis of our discussion. We have three requirements:

- We must be able to create a digital form of the image in the computer.
- We need appropriate software for processing the image.
- We need a computer which can perform the processing in an acceptable period of time with the required resolution.

Many comments here are similar to those we made in discussing the microscope itself. For example, you may have to work with the built-in system or the system that's already available in the lab. The difference is that some of the freeware programs are extremely powerful, so that all you need is the desktop computer. Many programs designed for desktop publishing are relatively inexpensive.

Thus, you can almost always find a way to extend your processing capabilities.

The motivation is that we need to obtain more information from an image than we can get by just looking at it. This principle ap. lies to more than HRTEM; we are discussing it here because HRTEM is where at present it is most needed/used in TEM. However, any TEM, X-ray map, or energy-filtered image or DP may benefit from processing and quantification. We need to quantify the TEM parameters, in particular C_s. One unique aspect of image processing in the TEM is that we have a choice between on-line and off-line processing. In fact, we often use on-line processing (frame averaging and background subtraction on the video image) to see the image even though the image we record may be unprocessed.

30.3. A CAUTIONARY NOTE

For most of our discussion we will consider only processing techniques using computers. To a large extent we can simulate a TEM using the computer. As we saw in Chapter 29, we can model a crystal, insert apertures, define the electron beam, including its broadening in the specimen, and then calculate the image. What we do in image processing is start with the image, add apertures and special filters, and then create a new image, the processed image. This image is a *real* image. What we must be careful about is explaining just what processing procedures we have used, since these may affect the interpretation of the data. This reporting is particularly critical when the raw data (the "original" image) are not being reported at the same time.

> Always report how you have treated your image, so that the reader can compare your data with related data that may have been processed differently or not at all.

30.4. IMAGE INPUT

There are several methods you can use to put the TEM image on the computer. The choice depends in part on how much detail you want in your digitized image, but also depends on how much work you're prepared to do. In this discussion, we'll only consider images which you have looked at on a video monitor, a CRT, or the fluorescent screen. Your basic choices are:

- Transfer the image directly from the TEM to the computer.
- Record the image on film, then digitize it using a microdensitometer.
- Record the image on video tape.
- Record the image on film, then print it and use a flat-bed scanner.

There are many methods for creating a digital form of an image in your computer. The simplest is to use a slow-scan CCD camera, which we discussed in Chapter 7. The drawback to CCD cameras is that high-quality CCD chips are very expensive for 1k × 1k arrays and astronomically expensive for 2k × 2k arrays. Although such cameras may become routine add-ons for all TEMs in the near future, even when you do have such a camera you will probably also want to use film or video. With film, you can record a larger area than you can using a CCD; you should use a video-recorder for *in situ* studies when using a heating or straining holder.

We can transfer the image from a video tape or a video camera to the computer using a frame grabber. Frame-grabber boards are readily available for most computers. You can use a high-resolution scanner for photographs or negatives. At this time, scanners with a resolution of 1k × 1k cost about the same as a video camera with comparable resolution. The purist's approach is to use a microdensitometer to measure the intensity of the film point by point and read this directly into the computer. The advantage of the microdensitometer is that it is very precise and can achieve the highest resolution for a very large area. The main problem is that it is slow, being a serial-collection technique. If you use it to its best advantage, your image will require a large amount of computer memory, which in itself is not a problem, but manipulating such images will still be slow.

30.5. PROCESSING TECHNIQUES

30.5.A. Fourier Filtering and Reconstruction

The principle involved in filtering is that a mask is used to remove some information from an image in order to enhance or emphasize other information. As an extra complication we can process the image, e.g., Fourier transform an HRTEM image, then apply a mask and then reverse the processing.

We can vary the size of the apertures and the sharpness of their edges. This is not possible in a modern TEM with normal fixed-diameter objective apertures. A single

30 ■ QUANTIFYING AND PROCESSING HRTEM IMAGES

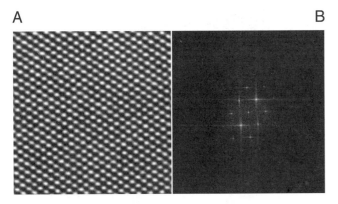

Figure 30.1. A square mask has been used to select the area shown in (A) from a much larger print of the image. The Fourier transform of this region is shown in (B) where you can see not only the spots in the 110 DP but also long streaks which run normal to the edges of the mask.

variable SAD aperture was used on some early TEMs, but it was triangular in shape and used three movable blades, so it was not much use as an objective aperture. You can best understand the procedure by an example. A square mask was used to select the region in Figure 30.1A from a much larger region of the HRTEM image, and its Fourier

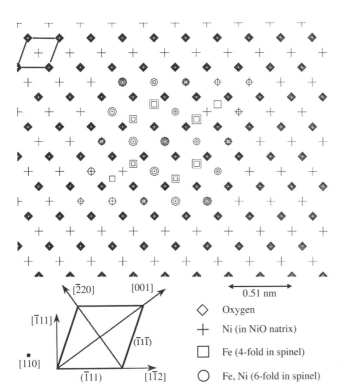

Figure 30.2. A model of an octahedron of spinel which is fully enclosed within a matrix of NiO. The rest of the specimen could then be modeled by adding extra layers of NiO above or below the defect layer.

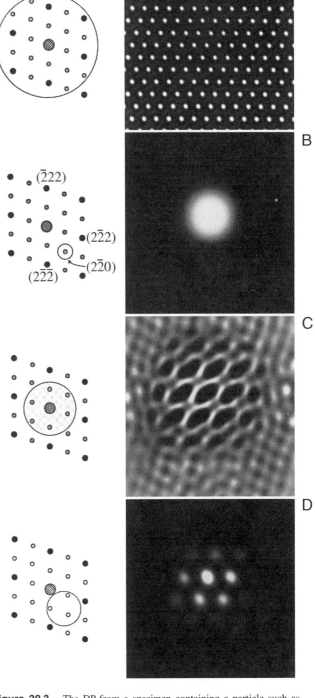

Figure 30.3. The DP from a specimen containing a particle such as that shown in Figure 30.2 is shown schematically in (A) together with the resulting lattice image. The other three pairs of diagrams illustrate how we can use the computer to produce different masks and thus generate different images, such as the DF image in (B). The image in (C) corresponds to the image shown in Figure 28.16, while the DF lattice image shown in (D) is analogous to that discussed in Figures 28.14 and 28.15.

transform (i.e., DP effectively from a few nanometers) is shown in Figure 30.1B.

What this technique does is to allow you to do microscopy in the computer. Your image becomes the specimen. You form the DP, then you can use apertures to select one or more beams to form the image; these apertures are the computer version of the objective aperture. Small apertures limit resolution as in the "real" TEM because the information about anything other than the perfect lattice is carried between the reciprocal lattice points. Figure 30.2 gives an illustration of how a model can be constructed of a particle in a matrix, which can be useful when simulating HRTEM and conventional BF/DF images in the computer, as shown in Figure 30.3. This was done using the Digital Micrograph package (Section 1.5).

30.5.B. Analyzing Diffractograms

In Chapter 28 we showed that the transfer function could be plotted out as shown schematically in, e.g., Figure 28.4. Another way of thinking about this plot is to imagine what would happen if you had a specimen which generated equally every possible value of **u**, i.e., every possible spatial frequency.

An amorphous film of Ge can provide just such a plot, but it is difficult to record the result because the scattered intensity is low.

> You will often see similar diffractograms obtained using a film of amorphous carbon. While such films are easier to make, they give little diffracted intensity for the range of **u** values between 6 and 8.5 nm⁻¹, which is important in HRTEM.

We thus record the image at high resolution, preferably directly using a slow-scan CCD camera, although digitizing the negative is fine. Then, by comparing your experimental plot of I versus **u** with those calculated for different values of Δf and C_s you can determine the astigmatism, the defocus, Δf, and the value of C_s (as we'll see below). It helps if you have a few particles of Au on the Ge film since the Au spots then give an internal calibration. Such a set of images and their corresponding diffractograms is shown in Figure 30.4. Notice that as the defocus of the objective lens increases, the number of rings increases but the rings become narrower. The contrast transfer gradually extends to larger values of **u**.

Determining astigmatism. You can use such diffractograms to correct the astigmatism, since a perfectly stigmated image will give a DP with circular symmetry. As you can see in Figure 30.5, even a small amount of astig-

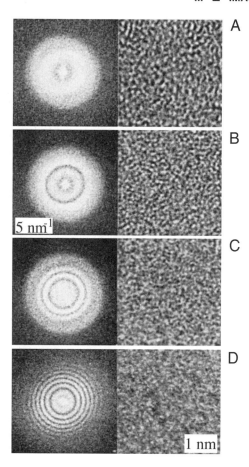

Figure 30.4. Four images of an amorphous Ge film and their corresponding diffractograms. Δf has the following values: (A) 1 sch; (B) −1.87 sch; (C) −2.35 sch; (D) −3.87 sch.

matism can be detected by eye. The computer can readily measure and correct such a pattern, as we will see shortly. This set of diffractograms also shows that the computer can distinguish astigmatism and drift in the image while your eye can easily mistake one for the other. Drift produces a circular pattern but the higher spatial frequencies are lost in the direction of drift.

Determining Δf and C_s. You can determine Δf for any image by measuring the radii of the bright and dark rings in the diffractogram, since bright rings correspond to $\sin \chi(\mathbf{u}) = 1$ and dark rings correspond to $\sin \chi(\mathbf{u}) = 0$.

$$\sin \chi(\mathbf{u}) = 1 \quad \text{when} \chi(\mathbf{u}) = \frac{n\pi}{2} \text{ and } n \text{ is odd} \quad [30.1]$$

$$\sin \chi(\mathbf{u}) = 0 \quad \text{when} \chi(\mathbf{u}) = \frac{n\pi}{2} \text{ and } n \text{ is even} \quad [30.2]$$

Since C_s will also influence the location of the rings, you need at least two rings. Krivanek (1976) has given a simple

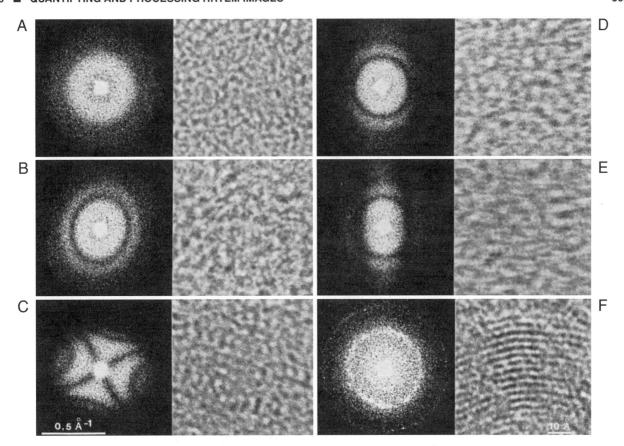

Figure 30.5. Six images of an amorphous carbon film and their corresponding diffractograms illustrating different misalignments of a 300-kV HRTEM: (A) well aligned and no drift; (B) some astigmatism (C_a = 14 nm); (C) more astigmatism (C_a = 80 nm); (D) no astigmatism but drifted 0.3 nm; (E) no astigmatism but drifted 0.5 nm; (F) well aligned and no drift showing graphite calibration fringes of 0.344-nm spacing. (B,D,E) $\Delta f = -2.24$ sch; (C) $\Delta f = 0$.

procedure for finding both C_s and Δf. If we start with our definition of χ

$$\chi(\mathbf{u}) = \pi \Delta f \lambda u^2 + \frac{1}{2}\pi C_s \lambda^3 u^4 \qquad [30.3]$$

then, inserting the values given in equations 30.1 and 30.2 leads to

$$\frac{n}{u^2} = C_s \lambda^3 u^2 + 2\Delta f \lambda \qquad [30.4]$$

All we now have to do is plot nu^{-2} versus u^2 to obtain a straight line with slope $C_s\lambda^3$, and with an intercept on the nu^{-2} axis of $2\Delta f \lambda$. Assign $n = 1$ to the intensity maximum of the central bright ring, $n = 2$ to the first dark ring, etc. The analysis can be trickier if you have used an underfocus condition or if you are very close to Scherzer defocus, but you will know when you have not found a straight line. You should find that your value of C_s will be close to that given by the manufacturer! A rather neat result is that, if you plot nu^{-2} versus u^2 for different diffractograms (i.e., different values of Δf), then the points corresponding to each particular value of n will lie on a hyperbola, as shown in Figure 30.6A. You can use these hyperbolas to determine C_s for any microscope and Δf for any diffractogram (Krivanek 1976).

Diffractograms and beam tilt. Beam tilt is very difficult to correct by eye; even worse, it can cause the diffractogram to look astigmatic so you correct the astigmatism instead. In the image, as we saw earlier, beam tilt can improve the appearance but confuse the interpretation! The set of diffractograms shown in Figure 30.6B shows you how to overcome the problem. You have to compare diffractograms taken at different beam tilts to determine the zero-tilt condition. A pair of diffractograms taken at $\pm \theta°$ tilt will only look the same (though rotated) if the beam had zero tilt at $\theta = 0°$. In the example shown, the diffractograms above and below the horizontal line are similar, so θ_y was very close to zero for the central condition. However, the pairs of diffractograms on opposite sides of the vertical axis differ slightly, so the alignment of θ_x was not perfect.

30.5.C. Averaging Images and Other Techniques

If you have recorded a series of images using a video camera, for example, you can average them over several frames as your eye does automatically. The result of such a process is illustrated in Figure 30.7. Different methods can be used to average the images. The easiest approach appears to be as good as any and simply involves taking the unweighted average of your best images, i.e., in the video example, just average over a series of frames. If you know that the object you're studying has a certain symmetry, you can use that information to improve the image further. The article by Trus *et al.* (1992) will give you a start on this process. If you want to remove the blur due to motion of the image, then you will really need to delve much more into this subject.

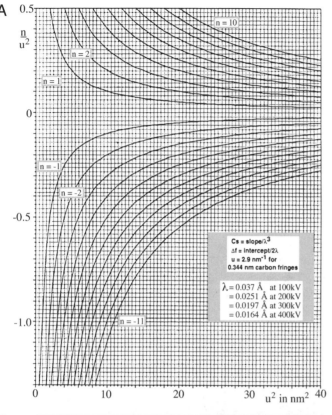

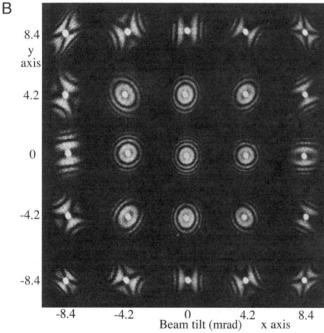

Figure 30.6. (A) Plot of nu^{-2} versus u^2. The rings in any diffractogram correspond to a series of n values which allow you to draw straight lines on this figure and thus determine the slope and the ordinate intercept, giving C_s and Δf respectively. (B) Set of diffractograms showing the effect of incident beam tilt.

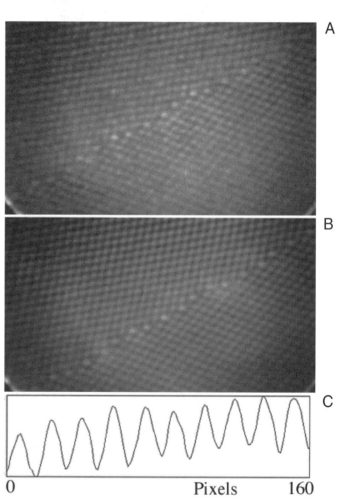

Figure 30.7. An example of the benefit of frame averaging to improve information from a video recording: (A) one frame, (B) 16 frames, (C) intensity profile along a (111) plane in (B).

If you use a TV-rate video, you'll almost certainly use background subtraction as a matter of routine. For example, you can record an image of the honeycomb pattern of the YAG detector, store it, and then automatically subtract it from all subsequent images in real time.

You may find it useful to add artificial color (pseudo-color) to your TEM images. Although it is often assumed that this is just done to make the images even more appealing to the nonmicroscopist (or nonscientist), there is actually a valid reason for the practice. Our eyes are much more sensitive to small variations in color than they are to small variations in gray level. You might therefore find color useful if you have a wide range of gray levels and want the viewer to be able to "see" some subtle variations. Similarly, you can use color to emphasize a particular gray level in an image. However, you have to be very careful in your choice of look-up table (LUT), the table which relates each gray level to a particular color. To get a feel for the dangers, play with Photoshop™ and your favorite TEM image. In the TEM, all of our apertures have relatively sharp edges. In the computer, you have the possibility of using multiple apertures, apertures with different shapes, and apertures with diffuse edges. Apertures with diffuse edges will help eliminate the streaking which will otherwise be present. You can also use the computer to do "unsharp masking," which is not the same as simply using a diffuse mask. The technique comes from the photographic process whereby we first print and image out of focus onto film, thus making a complementary image, except where there is fine detail present in the original image; in digital processing this is called Laplacian filtering. Many more examples are given in Russ's two books (Russ 1990, 1995).

30.5.D. Kernels

A kernel is simply an array of numbers which we can use to perform operations on a digital image. If we have the 3×3 kernel, K (we can have 5×5, 7×7, etc. but the computation time becomes too long, especially for real-time situations)

$$K = \begin{matrix} -1 & -1 & -1 \\ -1 & +8 & -1 \\ -1 & -1 & -1 \end{matrix} \qquad K_o = \begin{matrix} A & B & C \\ D & E & F \\ G & H & I \end{matrix}$$

we can apply it to every 3×3 group of pixels in our image, eg., K_o, and put the result in a new digital image. If we call our new 3×3 image K_i, then

$$K_i = \begin{matrix} A' & B' & C' \\ D' & E' & F' \\ G' & H' & I' \end{matrix}$$

The new image will have, for example, $E' = 8E - A - D - G - B - H - C - F - I$. This kernel then gives us a digital Laplacian (an approximation to the second linear derivative, ∇^2). What this kernel is doing is subtracting the brightness value of each neighboring pixel from the center pixel. If the area is a uniform gray, it will become white, so changes in contrast will be exaggerated. We can design a wide range of kernel operators. For example, the edge enhancer kernel has the effect of digitally differentiating the image. (We'll see a related digital-processing procedure applied to spectra in Chapters 35 and 39.) The Sobel and Kirsch operators are examples of such edge detectors; each can be thought of as the sum of several kernel operators. We can also use binary morphological operators which make binary features become larger or smaller. All of these operations can be carried out in any standard image processing package. In general, you should be very careful when using such techniques in TEM; their value is in displaying data which might otherwise be missed, rather than helping you quantify an image.

30.6. APPLICATIONS

This section will give you a taste of how image processing is being used now. It is just part of a rapidly growing list, so we are not going to be detailed or inclusive. We can separate the applications into two groups:

- Noise reduction or improving the signal/noise ratio.
- Quantifying images.

Of course, the first topic is included in the second.

30.6.A. Beam-Sensitive Materials

Low-dose microscopy necessarily implies that the signal-to-noise ratio will not be large; if it is large, the dose could have been smaller. This problem has been extensively addressed in biological EM and led to Klug's Nobel Prize for "Development of crystallographic electron microscopy and the structural elucidation of biologically important nucleic acid–protein complexes" in 1982 (see, e.g., Erickson and Klug 1971). In materials science we have tended to accept "beam damage" as a fact of life, but this attitude will not be acceptable for future quantitative HRTEM. Most modern microscopes will allow you to perform all your alignments on one area and then translate the beam a pre-

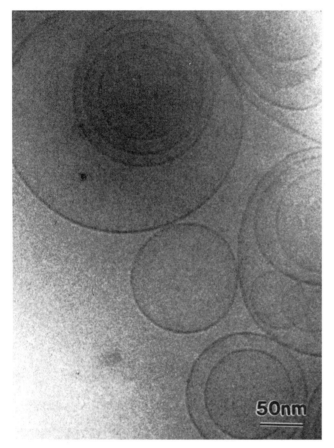

Figure 30.8. The image is from a highly beam-sensitive solution of surfactants in water. The solution has been frozen by plunging a film into liquid ethane and then transferring it to the TEM. The large circles are the surfactants that have aggregated to form vesicles; the concentration of the surfactants in the solution is just right for them to form lamellar structures. Texture starts to appear in the image as soon as the beam interacts with the specimen.

determined distance in a predetermined direction before recording the image of a pristine area. Clearly, the CCD camera will not only let you see your image without waiting to develop the plates, but you can take a series of images for noise reduction purposes and/or assess whether the imaging conditions were what you had intended. The image shown in Figure 30.8 illustrates the possibilities. If you read the review by van Heel *et al.* (1992), you will get some idea of how far you can already go in this field.

30.6.B. Periodic Images

In discussing quantitative analysis, we have already noted how we can use the computer to identify similar features and combine them in order to reduce the noise. This technique has many possible variations. Again, biological applications are leading the way with 3D crystallographic reconstruction (Downing 1992, Dorset 1995) and even correcting for distortions in the specimen (Saxton 1992).

30.6.C. Correcting Drift

Although drift is not as limiting on new machines, many older TEMs are still in use. Drift can be corrected now if the rate and direction of movement are constant. The computer can calculate the relative translation of two images and change the current in the image translation coils appropriately (which avoids moving the specimen). The difficulty is that the drift may not be linear. When implemented, such routines will be particularly valuable for frame averaging using a video camera. Then there will also be many applications for diffraction-contrast imaging, as well as for microanalysis.

30.6.D. Reconstructing the Phase

Although we are studying phase contrast, the image intensity doesn't directly give us phase information. Kirkland *et al.* (1982) have shown that the phase can be reconstructed by processing a defocus series. In their approach they use an iterative nonlinear image-processing technique to reconstruct the complex electron-transmission function. The technique was demonstrated using images of $CuCl_{16}PC$ (hexadecachlorophthalocyanine copper).

Five images from the experimental defocus series are shown in Figures 30.9a–e, together with the reconstructed transmission function plotted both as a real and imaginary part and then as an amplitude and phase part Figures 30.9f–i. The projected structure of the known unit cell is also shown Figure 30.9j. The phase image contains most of the structural information: it corresponds to the projected potential while the amplitude image contains features due to inelastic scattering. Notice in particular that we can now identify the benzene ring. This is one of the few examples of full phase reconstructions published. You must record such a series of defocus images if you want to do quantitative HRTEM.

30.6.E. Diffraction Patterns

We have generally ignored the intensities in DPs because they are so strongly influenced by dynamical scattering. However, if the specimen is very thin, we can use the intensity in the SAD pattern to carry out electron crystallography in the same way as in classical X-ray crystallography. As you can appreciate from Figure 30.10, particularly if the unit cell is large and the specimen examined is thin, there is a great deal of information in the SAD pattern but

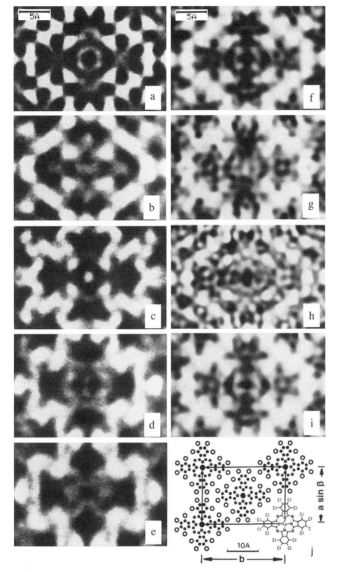

Figure 30.9. (a–e) Images from an experimental defocus series of $CuCl_{16}PC$; the reconstructed transmission function plotted as both a real and imaginary part (f,g) and then as an amplitude and phase part (h,i); (j) the projected structure of the crystal.

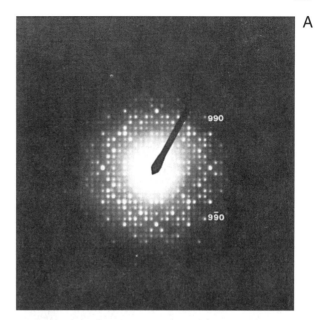

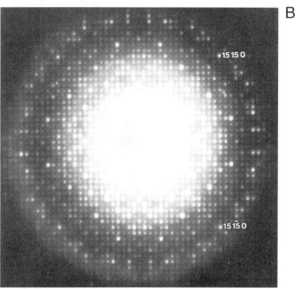

Figure 30.10. SAD patterns from $K_2O.7Nb_2O_5$ recorded using two different exposures. More than one exposure is needed to get all the information in the pattern. The space group in P4bm with $a = b = 2.75$ nm. The (15, 15, 0) reflections correspond to spacing of 0.13 nm.

you can't get it all in one exposure. Hovmöller's group (see, e.g., Zou 1995 and the ELD program in Section 1.5) have provided a routine for analyzing such patterns and getting structure-factor information. For example:

- Several patterns are recorded using exposure times of 0.5 s to >15 s.
- The patterns are digitized directly from the negatives using a CCD camera and a light box for backlighting.
- The intensity of each film is calibrated using a calibration strip with 20 equal exposure steps.
- The intensity is measured for all the spots and the processing begins.

This digitization process is particularly demanding, because each reflection typically covers an area of < 0.5 mm diameter on the photographic film. You will need to be able to index three strong, but clear, reflections. The computer can then perform a series of functions:

- Optimize the location of these points using a center-of-gravity approach, locate the origin, and index the rest of the pattern.
- Extract the intensities of each peak, taking care not to be misled by any shape effects of the specimen.
- Use reflections which are present on two successive negatives (since the intensities are now in digital form) to calibrate films recorded with different exposures and thus develop a very large dynamic range.

A cooled slow-scan CCD camera will give you a large dynamic range and better linearity than its room-temperature counterpart and should simplify this type of analysis. There are other complications in using electrons rather than X-rays for this kind of crystallography. While the Ewald sphere is still curved, as with X-rays, electrons can easily damage your specimen. However, the technique clearly has potential! Like all TEM techniques it can be applied to much smaller regions of the specimen than is possible for X-ray beams. We can also use the symmetry present in the SAD patterns. Because the specimen is very thin, this technique could be described as "kinematical" crystallography and complements the "dynamical" electron crystallography that we described for CBED patterns from thicker specimens in Chapter 21. The process of extracting intensities from DPs can now be carried out using the ELD software package (Section 1.5).

> The structure deduced from the HRTEM approach should generate the experimental SAD pattern, so it should be possible to use these diffraction data to further refine the structure.

If we can use the quantitative information available in DPs, we could combine this information with our experimental and simulated HRTEM images. The quantitative analysis of the DP is known as structure-factor-modulus restoration or reconstruction (Tang *et al.* 1995). One limitation of this approach is that the specimen be sufficiently thin that diffraction is kinematical. Of course, this requirement is necessarily similar to the HRTEM requirement of the WPOA.

30.6.F. Tilted-Beam Series

Having gone to great trouble to remove any beam tilt, we will now mention how beam tilt can be used to extend the resolution of your microscope! The basic idea goes back to the tilted-beam lattice-fringe imaging we discussed in Section 27.2. Now you use a computer to combine information in different tilted-beam images. The method proposed by Kirkland *et al.* (1995) assumes that you know when the beam tilt is zero. You tilt the beam through different angles in well-defined directions so that you transfer information in overlapping regions of reciprocal space, as shown in Figure 30.11A; you also need the on-axis image, as shown in the tableau in Figure 30.11B. Since it is important that

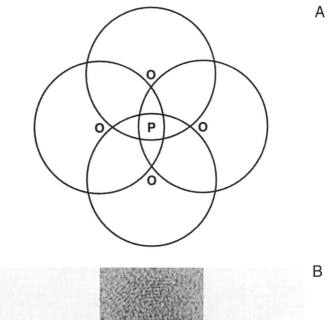

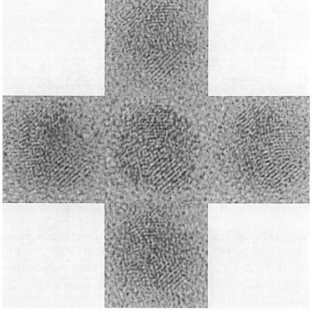

Figure 30.11. A method for extending the resolution of your TEM. Set the beam tilt to zero, then tilt the beam through different angles. (A) The four regions of Fourier space are shown by the four circles; O is each position of the tilted beam, P is the optic axis, and PO corresponds to the angle of tilt. (B) The five images used in the restoration arranged according to the beam tilt used in (A) with the on-axis image at the center.

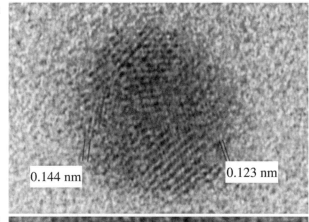

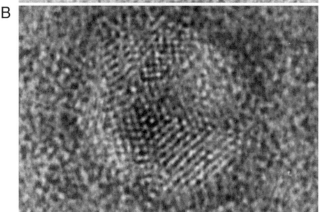

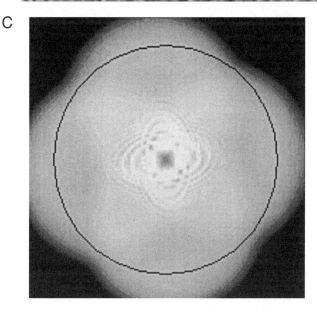

Figure 30.12. The restored image of a gold particle on amorphous Ge: (A) the amplitude (modulus) image showing 0.123-nm fringes, (B) the corrsponding phase image, (C) the transfer function after restoration, plotted in two dimensions. The circle corresponds to 0.125-nm detail successfully transferred to the image. For such thin specimens, atom-dense positions have a reduced modulus [black in (A)] and an increased phase [white in (B)].

the same area is imaged, a sixth (on-axis) image is recorded and correlated with the first on-axis image to check for drift and specimen degradation. You now need to restore the modulus and phase to create a higher-resolution image.

Kirkland's paper is a beautiful demonstration of the care needed in image processing. Even aligning the images is not trivial. However, the resulting restoration shown in Figure 30.12 demonstrates the potential of the technique: detail is present in the image at a resolution of 0.123 nm using a 400-kV microscope.

30.7. AUTOMATED ALIGNMENT

In the not-too-distant future, all TEMs will have automatic beam alignment, astigmatism correction, and readout for Δf. What makes this possible is the diffractogram analysis, a slow-scan CCD camera to digitize the image, and computer control of all the microscope functions. By microscope functions we mean all lens currents, deflector currents, specimen drive, and aperture drives. The slow-scan camera is needed because the computer needs to make measurements on more than one ring in the diffractogram. So you will not actually have to sit in front of the microscope once you have loaded the specimen.

> The big advantage in remote control will not be just that you can sit in Bethlehem and operate a microscope in California but that you can locate the microscope in its own controlled environment with no one opening the door to check if it is working (it was) or entering the room and thus changing the heat-load. Also, if your specimen is not ideal or the microscope breaks down, you won't have to go for a walk on the beach in California but can continue word processing in Pennsylvania.

The procedure has been extensively discussed and implemented by Saxton and Koch (1982), Krivanek and Mooney (1993), Koster *et al.* (1988), Koster and de Ruijter (1992), and others. Your role is to select a suitable region of the specimen close to the area of interest; the area of interest should ideally only be examined at low magnification. At present, you will make the initial alignment manually and then turn the process over to the computer. The computer will then adjust the astigmatism and correct the beam tilt independently and quickly.

Figure 30.13 shows how well and quickly this procedure can now be done. The different diffractograms in each tableau correspond to incremental changes in the

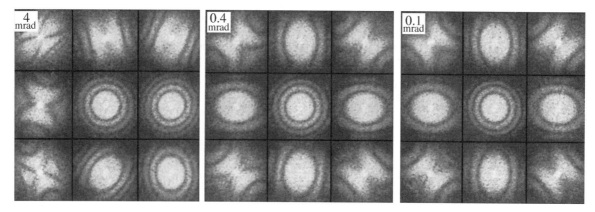

Figure 30.13. Using the computer to correct the beam tilt: The different diffractograms in each tableau correspond to incremental changes in the beam tilt of 6 mrad in the x and y directions away from the initial beam tilt in the central diffractogram. The computer initially determined a misalignment of 4 mrad, then corrected this to 0.4 mrad and finally to 0.1 mrad. Note the central diffractogram is almost unchanged, emphasizing the need for computer-controlled tilting to give correct alignment.

beam tilt of 6 mrad in the x and y directions. The computer showed that the initial tilt error was 4 mrad, which was reduced to 0.4 mrad after one pass and < 0.1 mrad after the second pass. Each pass only took 28 s! The astigmatism shown in Figure 30.14 was initially 53 nm. It was reduced to 3 nm after one pass and to < 1 nm after the second pass. For this correction each pass took only 8 s. Even an experienced operator can't match this speed or accuracy for either correction, and both corrections are now *quantitative*.

The defocus value is then found by calibrating the image with minimum contrast at Δf_{MC}. The value of Δf_{sch} can then be found when the image contrast is a maximum. Although the method described here uses the diffractogram, a corresponding approach can be followed by analyzing variations in the contrast of the image. This technique has been described by Saxton *et al.* (1983) and uses a method of cross-correlating pairs of images recorded at each focus setting of the microscope. The reason for cross-correlating images is to remove the effects of electron shot noise; variations due to the photographic emulsion are avoided by using the slow-scan camera.

30.8. QUANTITATIVE METHODS OF IMAGE ANALYSIS

In the next six sections we will go through several particular illustrations of image processing in HRTEM. Our discussion will draw heavily on the work of a few pioneers in this field; we will also emphasize that, although this subject is still in its infancy, it is developing rapidly. The main cause for the delay in its application in materials science has been the lack of affordable fast computers and the feeling that everyone must write their own image processing program; the latter is not true and is certainly not recom-

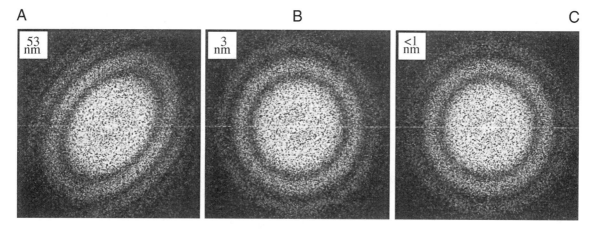

Figure 30.14. Diffractograms showing the astigmatism corrections made by the computer following a similar procedure to that shown in Figure 30.13. The final diffractogram shows that the HRTEM is now very well stigmated.

mended. At this time we can summarize the situation as follows:

- Quantitative analysis is difficult, often tedious, and invariably time-consuming.
- You have to understand the basic ideas of image theory.
- Your analysis is only as good as your image and your image is only as good as your specimen.

We gave the necessary information on how to obtain the software in Section 1.5.

30.9. PATTERN RECOGNITION IN HRTEM

The most obvious feature of the majority of HRTEM images is that we see patterns of white, gray, and black dots or other shapes. If the pattern is perfect everywhere, your specimen is probably a single crystal with no defects, no thickness variations, no variation in atomic composition, and no use. If it is not perfect, then we can use pattern recognition to quantify the variations.

> The principle of pattern recognition is to take or make a template, move it across your image, and measure how closely the image resembles your template.

Clearly you need a computer for this! Your template needs to match the magnification and rotation of the pattern you are examining. Then you need a method to say how close your match is, i.e., you need to know your "goodness of fit." We will go through some basics here, but strongly recommend that you consult the list of original papers given at the end of this chapter when you are ready to apply this technique.

We can illustrate the approach as shown in Figure 30.15 following Paciornik *et al.* (1996). The large rectangle represents your digitized image and could be 1k × 1k; remember, the numbers indicate pixels. The small rectangle represents your template. This template might be a small area of the pattern or a simulated image, in which case it might be a 128 × 128 pixel template. If the template is taken from your image, then you have already got the

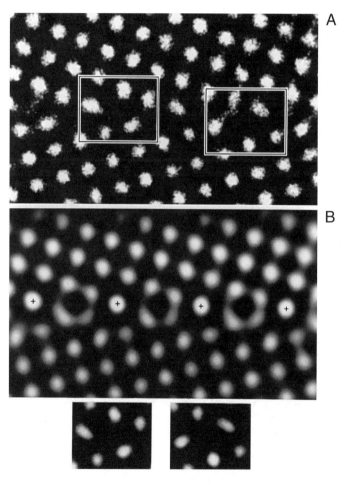

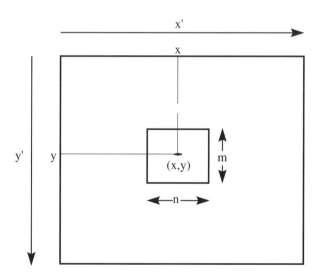

Figure 30.15. The large rectangle represents the digitized image, size $(x' \times y')$; the small rectangle, $(m \times n)$ pixels, represents the template used in the cross-correlation calculation. The small rectangle is moved to different (x,y) positions during the process.

Figure 30.16. Analysis of a small region of a $\Sigma = 5$ tilt boundary in TiO_2. The two small boxed regions in (A) are only present at the boundary; these are used as the templates for the cross-correlation method. (B) The cross-correlated image. The small rectangles at the bottom of the figure are the low-noise averaged images of the GB templates.

right magnification and rotation. If not, you have to set these first and we will return to this problem shortly.

> This is a real-space approach.

The process is best understood by an example. Figure 30.16A shows an HRTEM image of a small region of a $\Sigma = 5$ tilt boundary in TiO_2. The two small boxed regions appear only at the boundary and are selected as templates. The matching process has then been carried out and the new image is shown in Figure 30.16B. Having found all the regions which match the template, we could then take the average of these to produce low-noise images of the grain boundary templates. The final step is the comparison of these templates with models of the grain boundary structure. There are two important points to remember:

- When you average images, you implicitly assume that they are all the same.
- Don't forget our discussion in Chapter 27 of interface grooving and the problems associated with interfacial segregation.

30.10. PARAMETERIZING THE IMAGE USING QUANTITEM

In general, the thickness or chemistry will vary as you cross the specimen, i.e., the projected potential varies across the specimen. This means that one template will only match a small area, so you have to use many templates. These templates could, in principle, be totally empirical, but to be quantitative you must derive them from image simulations. This approach has been described for two special cases by Kisielowski *et al.* (1995) and Ourmazd *et al.* (1990).

30.10.A. The Example of a Specimen with Uniform Composition

In the QUANTITEM approach developed by Ourmazd, the results of Chapter 29 are summarized by a general equation linking the intensity and all the imaging (S_i) and materials (P) parameters,

$$I(x, y) = F(P(x, y), S_i) \quad [30.5]$$

This equation just tells us that the intensity depends on the imaging conditions and on the specimen. For a particular set of imaging conditions, S_i will be known (more or less) and we'll call it S_i^0. Then we can write that

$$I(x, y) = F\left(P(x, y), S_i^0\right) = F^0(P(x, y)) \quad [30.6]$$

The basis of this approach is quite straightforward:

- Define the function F^0 for each image that you may obtain.
- Then construct a set of templates for your matching process.

Providing you stay within one extinction band, F^0 will be directly related to the projected potential of the specimen. Ourmazd gives a helpful simple analogy for this process, as illustrated in Figures 30.17A and B. The function F^0 describes the path of a swinging pendulum as it varies with time. Each value of F^0 corresponds to a snapshot of the pendulum, so if you plot F^0 you can "see" the path of the pendulum. The velocity of the pendulum is related to the density of points along the path. So it should be possible to plot out the function F^0 from a single lattice image even if you don't know the microscope parameters used to form the image.

Yes, there are limitations and conditions and we'll discuss them later. All we need now is a method for representing each image by a snapshot of the pendulum: we have to *parameterize* the image. This process is the key to the technique. Manipulating and quantifying, in principle, thousands of images, each requiring 4 Mbyte of memory, is not a fast process, even if you do have that much memory. If we could characterize each image by a few numbers (a vector or parameter), the comparison process could be much faster.

We separate the image into unit cells and digitize these to give many templates, which are n pixels by m pixels as shown in Figures 30.17C–E. If we define N to be $n \times m$, then we have N numbers for the N pixels, where each number represents a gray level. Now the N numbers are regarded as the N components of an N-dimensional vector. (The math is not complicated, but don't try to visualize this vector.) So now all the information in each unit cell is represented by a vector in N-dimensional space. The function F^0 describes the path of these N-dimensional vectors as the projected potential changes.

The next step is to define a reference frame for these vectors. Three basis vectors are derived from the experimental image. Ourmazd *et al.* (1990) argue that three basis vectors will be sufficient, as we can show in the following way. We will be using a low-index zone axis for any HRTEM analysis. Then we have three types of images:

- The background, $\mathbf{R}^B$, due to the direct beam, O.
- A single-period image, $\mathbf{R}^S$, due to the interference between O and the strongest reflections, G_i.

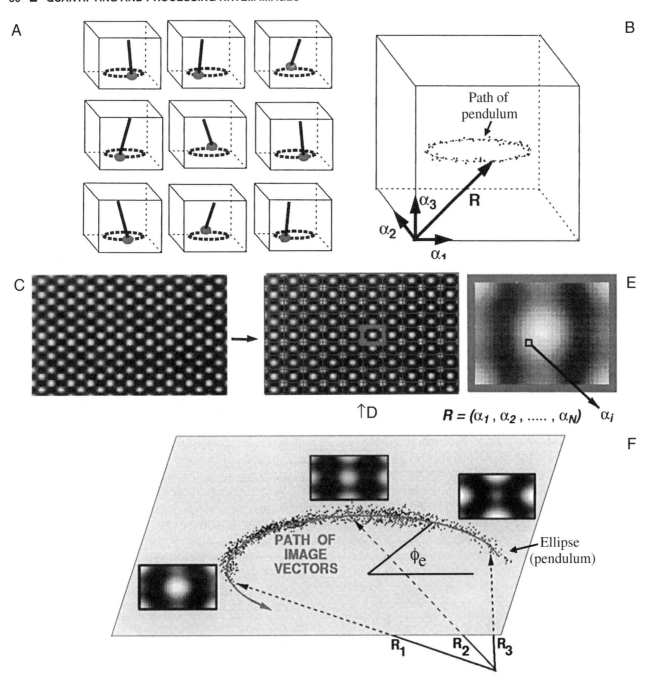

Figure 30.17. The principle of vector parameterization used in QUANTITEM. (A) shows the different pendulum positions and (B) shows the path of the pendulum. Each HRTEM image is represented by a single vector R which has N dimensions; (C–E) The image is separated into unit cells and digitized to give $(n \times m)$ pixel templates; (F) Three vector-parameterized image ($\mathbf{R}_1$, $\mathbf{R}_2$, $\mathbf{R}_3$) of a wedge-shaped specimen of Si at different thicknesses.

- A double-period image, $\mathbf{R}^D$, due to the interference between these strong G_i reflections.

Each of these **R** terms is a vector which represents an image. Any image we can form must be a combination of these three types of image, so a general image, G, can be written as

$$\mathbf{R}^G = a_G \mathbf{R}^B + b_G \mathbf{R}^S + c_G \mathbf{R}^D \quad [30.7]$$

Each of the basis vectors (images) can be expressed in the same manner

$$\mathbf{R}_i^T = a_i \mathbf{R}^B + b_i \mathbf{R}^S + c_i \mathbf{R}^D \quad [30.8]$$

giving three vectors for $i = 1, 2,$ and 3.

We can turn these equations around to define any vector $\mathbf{R}^G$ in terms of the basis vectors

$$\mathbf{R}^G = \alpha_G \mathbf{R}_1^T + \beta_G \mathbf{R}_2^T + \gamma_G \mathbf{R}_3^T \quad [30.9]$$

which is what we wanted to show.

Ourmazd *et al.* point out that this treatment gives three important results:

- The vector notation allows us to parameterize the lattice image.
- Projecting the vectors onto planes and/or paths aids noise reduction.
- Any noise which remains can be quantified.

The result of vector-parameterizing an experimental image of a wedge-shaped specimen of Si is shown in Figure 30.17F.

30.10.B. Calibrating the Path of R

In order to relate any image to the projected potential, we have to calibrate the curve showing the path of $\mathbf{R}^G$. This is where the image simulation comes in. We start with the vector-parameterized analysis of a series of simulated images as shown in Figure 30.17F. Each point on the curve corresponds to a unit-cell image, and thus to a vector $\mathbf{R}^G$. The ellipse has been fitted empirically, and the thickness of the cell has been increased by 0.38 nm for successive calculations. The points are closer together in some parts of the plot because, as we saw in Chapter 29, some characteristic images appear for a wider range of thicknesses. Now we have a way to quantify this "experimental" observation. What the ellipse does is to allow us to parameterize the path in terms of the phase angle of the ellipse ϕ_e, shown in Figure 30.17F. Thinking back to the pendulum analogy, the path parameters are the image version of the coordinates for the harmonic oscillator.

Now the ϕ_e curve parameters can be obtained from a series of images. We can change the material and in each case examine three other variables:

- The orientation of the specimen (i.e., the zone axis).
- The defocus, Δf, of the objective lens.
- The specimen thickness.

The remarkable result is that when we plot ϕ_e versus the thickness, normalized by the extinction distance, we obtain a straight line. The explanation for this result is related to the fact that only a small number of Bloch waves usually contribute to the image, as we saw in Chapter 29. In materials such as YBCO, this is not the case, because too many Bloch waves are important and the curve is not a straight line.

30.10.C. Noise Analysis

If noise moves the vector off the ellipse, we can analyze the noise. If it moves the vector exactly along the ellipse, we can't analyze the noise, but that is quite unlikely since the noise would then be accurately mimicking a change in projected potential. So this method should reduce the noise by a factor of $\sqrt{N}$, which for a 10 pixel × 10 pixel cell is a factor of 10!

The analysis given by Ourmazd *et al.* then shows that, in the case where only two Bloch waves are excited, the image intensity, I, can be expressed as

$$I = B + S + D \quad [30.10]$$

where B, S, and D are the contributions from the background, single interaction, and double interaction, as we defined them above. The point (B, S, D) does indeed describe an ellipse which lies on a plane independent of Δf.

The value of this approach can be appreciated if you look at the examples shown in Figures 30.18A and B. In the first example, the technique has been used to provide a map of the roughness of the Si surface. The experimental image looks really uniform until you analyze it using this method, when you can discern the roughness at the ~0.5-nm level, as in Figure 30.18B.

As you know from earlier discussions, changes in chemistry produce effects which are similar to changes in thickness, because they change the projected potential. In terms of the present analysis the effects are different: composition changes cause changes in the ellipse and in ξ (notice that there is no subscript, since this ξ is a many-beam value).

The method is more limited in respect to change in composition but can be used if the thickness and roughness are known, i.e., if you can measure the roughness elsewhere on your specimen (using a known reference cell), and infer it for the area you want to analyze. (Warning lights should be flashing.) The approach is as follows:

- Use QUANTITEM to measure the advance in ϕ_e at your target cell relative to your reference cell.
- Subtract $\Delta \phi_e$, which is due to a thickness change.
- Then the rest of the change in ϕ_e must be due to changes in ξ. If you know how ξ varies for dif-

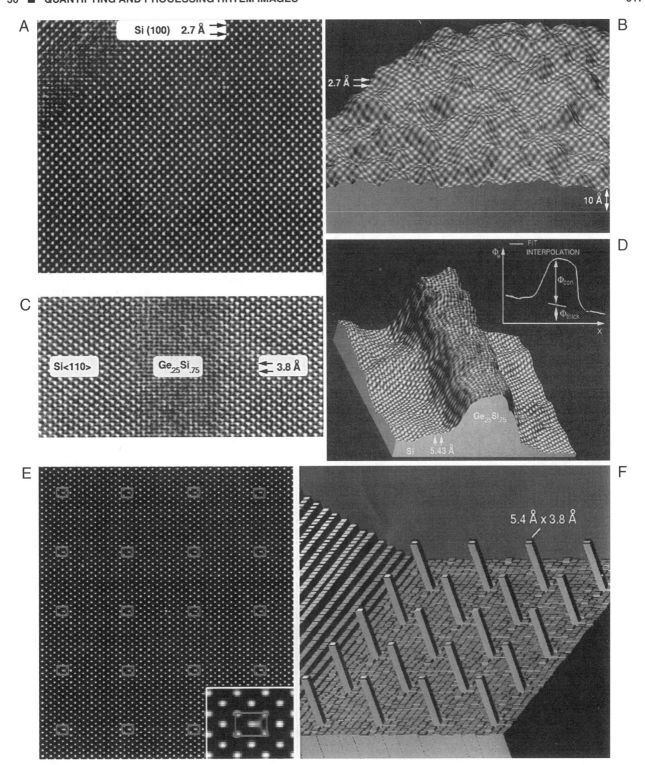

Figure 30.18. Examples of the application of QUANTITEM: on the left is the conventional HRTEM image, on the right is the QUANTITEM image. (A,B) Mapping of the roughness of the Si surface covered by SiO_2; (C,D) a layer of Ge_xSi_{1-x} in a matrix of Si, the inset shows the plot of ϕ_e versus x; (E,F) analyzing a simulated image of columns of Ge (a δ-function in concentration) in Si.

ferent compositions, you have determined the local composition.

The example shown in Figures 30.18C and D is a near-perfect application for the technique, since the elements in the alloy Ge_xSi_{1-x} are randomly located on the lattice sites. The slope tells us how abruptly the composition varies.

You can test the potential resolution of the technique, its sensitivity to the alignment of the beam, bending of the specimen, and beam divergence in the usual way by creating model structures, simulating the images, and then analyzing them. Figures 30.18E and F show that the potential resolution is superb, but beam tilt can cause 10% errors in thickness measurement. The conclusion is clear: as always, you will only get the best results if the specimen is ideal and both the microscope and the specimen were perfectly aligned. Note, however, that the technique has not yet been successfully applied to a wide range of materials, but it is complementary in many ways to STEM Z-contrast (see Figure 22.15).

30.11. QUANTITATIVE CHEMICAL LATTICE IMAGING

This technique uses the approach described in Section 30.10, but can only be applied to materials where we have chemically sensitive reflections, which we discussed in Chapter 17. We will use these reflections in Chapter 31 to produce chemically sensitive DF images. In HRTEM, the chemically sensitive reflections not only contribute to the overall image but they will generally have a different dependence on thickness, too.

This effect is shown in Figures 30.19A and B for AlAs and GaAs, which have identical structures. The 002 reflection is allowed for both, but is stronger for AlAs since F, the structure factor, is proportional to $f_{III} - f_V$. You can see that, under the conditions chosen for this comparison, the intensity of the 022 reflections is also very different for the different thicknesses. Figure 30.19C shows the sort of image we can analyze with this approach. We want to know how abruptly the composition changes at the interface.

In this example, the ideal GaAs and $Al_{0.4}Ga_{0.6}As$ unit-cell images are characterized by the two vectors, $\mathbf{R}_{GaAs}$ and $\mathbf{R}_{AlGaAs}$, following the approach described in Section 30.10. This was done in this case by simulating the cells, dividing them into 30×30 pixel arrays (so $N = 900$), and then plotting $\mathbf{R}$. The information content is contained in θ_C. As before, we can directly assess the noise in such an image. So how is the direction of $\mathbf{R}$ dependent on composition?

The technique is explained by Figure 30.19D. The three known simulated templates each produce a vector $\mathbf{R}^t$. Although the vector for the intermediate composition does not lie in the plane, it can be projected onto this plane to give a unique vector for certain ranges of thickness. Since this is a complex procedure, you'll find the flow chart shown in Figure 30.20 helpful.

- In Figure 30.20, the experimental image (A) is digitized; the image contained approximately 25×25 unit cells and used a 514×480 frame buffer.
- Next, the image must be separated into individual cells (B).
- The pair of templates shown (C) is then used to calculate the angular positions of the $\mathbf{R}$ vectors for all the unit cells. Such templates can be calculated or taken from known areas of the specimen.
- These $\mathbf{R}$ vectors are characterized in terms of where they cut through a plane (D) (see Figure 30.19D also).

The maximum chemical difference determines how far apart the two principal distributions can be (E). Since the image is now fully parameterized, we can do the statistics and finally invert the angular data to give the compositions (K).

This technique has enormous potential, but you must also remember that it is susceptible to all the drawbacks inherent in HRTEM. The big advance is that now you can put numbers on those effects. The technique is material-specific, but if you know your material, you can combine image simulation and this processing method to examine what will be the limiting factors for your material. You can construct a test image like that shown back in Figure 30.18E. If your specimen is ideal, you could, in principle, easily detect a column of Al in a mainly GaAs matrix without any "spreading" due to the electron beam.

30.12. METHODS OF MEASURING FIT

There are two methods presently used to obtain a measurement of the goodness of fit, namely:

- Cross-correlation.
- Least-squares refinement (Section 30.13).

In this section, we'll use the cross-correlation method to compare an $n \times m$ pixel template (see Section 30.9) with every possible $n \times m$ rectangle in the image. The computer

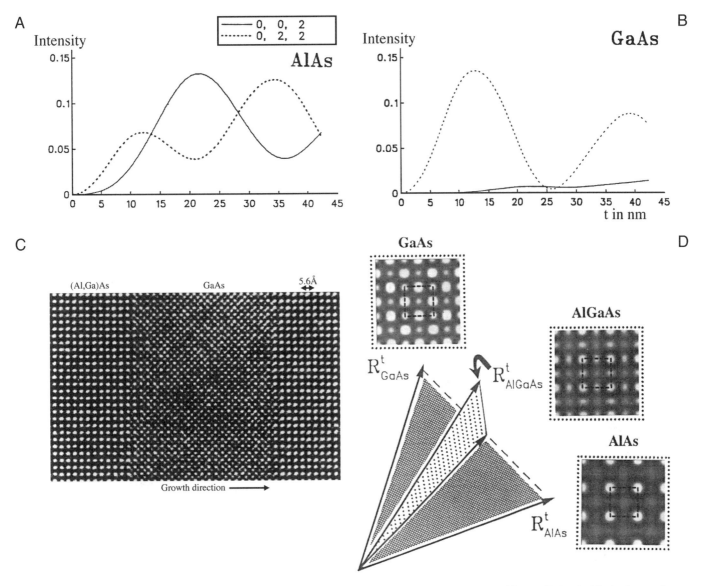

Figure 30.19. (A,B) Variation in intensity of the (002) and (022) beams along [100] in AlAs and GaAs (400 keV); (C) chemical lattice image of a layer of GaAs between two layers of AlxGa$_{1-x}$As ($x = 0.4$); (D) templates simulated for different values of x each produce a vector $\mathbf{R}^t$.

moves the template across the image one pixel column at a time, then shifts down one pixel row, and repeats the exercise. The cross-correlation function (CCF) gives the goodness of fit or a "measure of similarity" between the template and each $n \times m$ image

$$CCF(x,y) = \frac{\sum_{x'}\sum_{y'}[i(x',y') - \langle i(x',y')\rangle]\cdot[t(x'-x, y'-y) - \langle t\rangle]}{\sqrt{\left\{\sum_{x'}\sum_{y'}[i(x',y') - \langle i(x,y)\rangle]^2 \sum_{x'}\sum_{y'}[t(x'-x, y'-y) - \langle t\rangle]^2\right\}}}$$

[30.11]

In this equation x varies from 0 to x_{max}, y varies from 0 to y_{max}.

- $i(x', y')$ represents the image.
- $t(x', y')$ represents the template.
- $\langle t \rangle$ is the average value of the pixels in $t(x', y')$; it is computed just once.
- $\langle i(x, y)\rangle$ is the average of $i(x', y')$ in the region coincident with the current location of t.

The summations are taken over the coordinates common to both i and t. The origin of the image is at its top left corner and the origin of the template is at its center. In this equa-

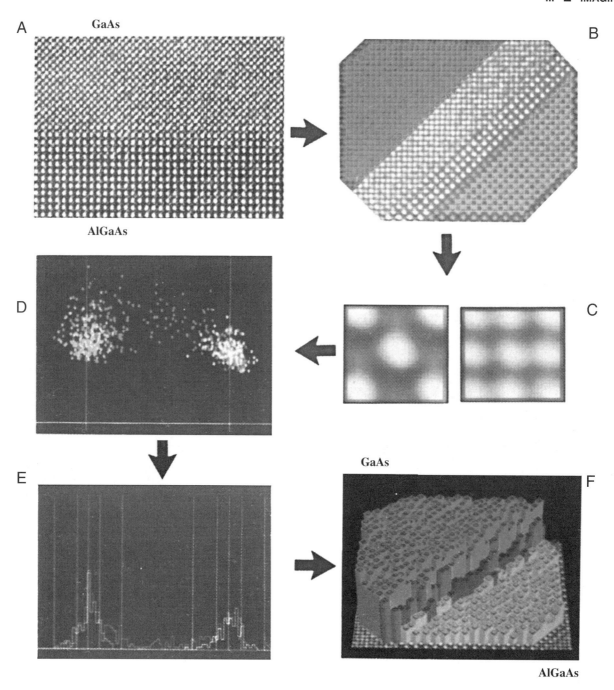

Figure 30.20. A flow chart summarizing the pattern recognition procedure.

tion the denominator is a normalization factor, so that the CCF will not depend on any difference in the intensity scale between the template and the image.

We can rewrite equation 30.11 as the dot product of two vectors **t** and **i** which gives us the $n \times m$ component of the template

$$CCF(x, y) = \cos(\theta) = \frac{\mathbf{t} \cdot \mathbf{i}}{|\mathbf{t}||\mathbf{i}|} \qquad [30.12]$$

Now we can plot the CCF as a map of our image and then examine it to deduce where there is a particularly good match. Since the CCF value varies from 0 to 1, we can plot

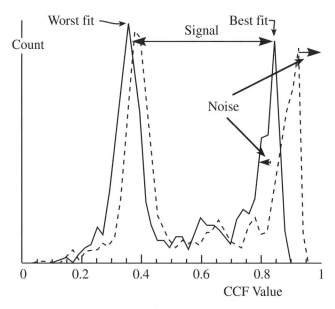

Figure 30.21. A plot of how often a particular CCF value occurs. The two peaks in the continuous curve are the best fit and the worst fit: their separation gives a measure of the discrimination signal; the width of the peaks gives a measure of the noise and hence a signal-to-noise ratio. The plot can be redrawn after repeating the process (dashed line) to estimate the improvement in signal-to-noise ratio.

out the number of times each particular CCF value occurs, as illustrated in Figure 30.21. The two peaks in this curve correspond to the best fit and the worst fit, so the distance between them gives a measure of the "discrimination signal." From the width of the peaks we have a measure of the noise, and hence a signal-to-noise ratio. The regions of good fit can be combined to produce a better template, and the process repeated, giving the dashed line. A second measure of the noise is then given by how far the good peak differs from unity. A particularly nice feature of this approach is that the procedure is available as a plug-in module for Digital Micrograph (see Section 1.5). Your template could alternatively be a simulated image and the process repeated for a series of different thicknesses and/or defocus values. When you want to learn more about correlation techniques, see the article by Frank (1980).

30.13. QUANTITATIVE COMPARISON OF SIMULATED AND EXPERIMENTAL HRTEM IMAGES

If we want to compare simulated and experiment images quantitatively, we really should modify our usual approaches to both simulation and experiment (King and Campbell 1993 and 1994). When doing the simulation, most programs automatically adjust the gray scale for each image so that darkest is 1 and brightest is 0 (or vice versa). This means that two simulated images might appear similar even though you would hardly see the pattern in one if both appeared on the same negative. In a similar way, we usually print an image to be as clear as possible using the full range of contrast of the photographic paper.

We need methods for normalizing these procedures if we want to make quantitative comparisons. The solution for the simulation is simple. For the experimentalist, it means recording extra data while you're at the microscope. After recording the image, you record another image with the specimen removed. You then use this image to scale your lattice image such that you correct for variations in intensity across the field of view and the nonlinearity of the response from the photographic film. Figure 30.22 illustrates the experimental transmittance for Kodak SO-163 film, 400-keV electrons, plotted against the digital value on a CCD array. Of course, you must process both images at the same time. This is called the "flat-field" correction; a slow-scan CCD camera would simplify this procedure at the cost of reducing the area you examine.

> Don't forget that since HRTEM uses higher voltages, the perfect image will only be recorded from an area of the specimen that has only seen the beam while you recorded the image! So you should always use low-dose techniques for quantitative imaging.

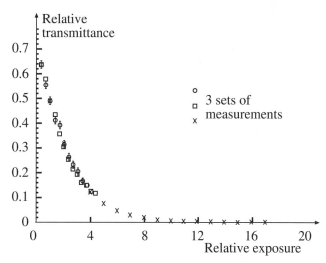

Figure 30.22. Plot of transmittance versus relative exposure measured using a CCD camera to digitize images from Kodak SO-163 film (the symbols indicate three different microscopes).

When you analyze the image, you'll find out if the area you photographed was correctly aligned. Since your image only takes two seconds or so to record, you may risk several exposures using this technique.

You are now comparing numbers, so you can use a least-squares fit where the residual $f_i(x)$ is defined as

$$f_i(x) = \frac{\left[f_i^{\text{obs}} - f_i^{\text{calc}}(x)\right]}{W_i} \quad [30.13]$$

and your task is to minimize $f_i(x)$. The difference between the intensity in the experimentally observed ith pixel and its calculated value would be zero if everything had been scaled correctly, the imaging conditions (Δf, C_s, etc.) were correct, and you have the right structure.

Let's say W_i is the image which represents the error bars for pixel i. Then we can write that

$$W_i = \min\left[\sum_{i=1}^{N} f_i(x)^2\right] \quad [30.14]$$

This equation defines the nonlinear least-squares problem. We use x to summarize a set of parameters (Δf, C_s, the model, etc.); N is the number of pixels in the image. Fortunately, this analysis is now routine statistics. You'll need a computer program to tell you how good the first guess was, make an improvement, and continue until it meets our specific criterion for matching [King and Campbell used MINIPACK-1 (Moré 1977, Moré et al. 1980)].

In their demonstrations of this approach to analyze a [001] tilt grain boundary in Nb, King and Campbell (1993, 1994) varied four parameters: thickness, defocus, x-tilt, and y-tilt. The steps were as follows:

- They first optimized the electron-optical parameters using a 64 × 64 pixel image, giving $N = 4096$ and an image computational cell of 3.303 nm by 3.303 nm. Using the EMS program (Section 1.5), the optimization took 20 iterations and 80 multislice calculations.
- Next, they had to optimize the structure of the grain boundary. This process required defining 84 atomic positions in a unit cell of 4.16 nm × 1.04 nm and a 512 × 128 (= 65,536) pixel image. Now the optimization required 16 iterations and 1300 multislice calculations.

These numbers are instructive. First they tell you that this computation can be done, which wasn't obvious. Secondly, they tell you that this is a computer-intensive process; that you could have guessed!

You'll need to take enormous care in this type of analysis:

- Align the simulated cell with the experimental cell and measure the unit cell in pixels.
- Choose a number of cells and relate them by the translation vector parallel to the rows of the image array.
- Calculate the standard-deviation images.
- Rotate the unit cell and repeat the exercise several times.

The orientation which gives the smallest standard deviation is your alignment. You must now adjust the magnification of the experimental image to fit the simulation, in a similar way to what you did for rotation. Next, you have to match the origins of both cells; the procedure is the same as we just described, but translating the unit cell, not rotating it. For a bicrystal, you now repeat this exercise for the other grain and then for the grain boundary. You can improve the fit further if you take account of a constant background contribution which probably arises due to the amorphous layer on both surfaces. Comparing experimental and calculated images quantitatively, we define f_i^{obs} as the intensity value of the ith pixel in the experimental image and f_i^{calc} as the corresponding value in the simulated image. We then calculate the residual $f_i(x)$ as follows

$$f_i(x) = \frac{\left(f_i^{\text{obs}}(x) - \left(f_i^{\text{calc}}(x) + b^{\text{fit}}\right)\right)}{W_i} \quad [30.15]$$

where b^{fit} is included as a free parameter in the optimization procedure. King and Campbell's calculations showed that W_i could be expressed as

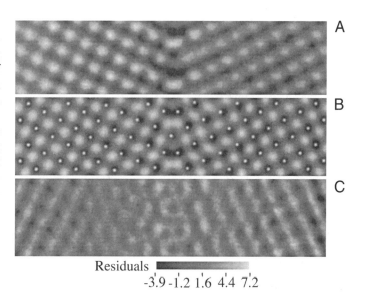

Residuals
-3.9 -1.2 1.6 4.4 7.2

Figure 30.23. (A) Experimental image, (B) best-fit simulation, and (C) normalized residuals of a $\Sigma = 5$ symmetric tilt boundary in Nb.

$$W_i = \sigma_i^{obs} + 0.05 f_i^{obs} \qquad [30.16]$$

where σ_i^{obs} is the standard deviation of the *i*th pixel. Examples of the experimental, best fit, and normalized residuals are shown in Figure 30.23 for images from a $\Sigma = 5$, (310), [001] GB in Nb. The values for the thickness and Δf show how consistent this technique can be, especially since the images in Figures 30.23A and B were from *opposite sides of the grain boundary*; (C) was like (B), but for a different defocus value.

30.14. A FOURIER TECHNIQUE FOR QUANTITATIVE ANALYSIS

Möbus *et al.* (1993) have proposed using what is referred to as an adaptive Fourier-filtering technique. The HRTEM image is digitized in the usual manner and then a special spatial-frequency filter is applied. This type of mask is most promising for analyzing regions which contain defects.

> An adaptive filter is one where the shape of the filter, or mask, is adapted to fit the shape of the "image" it's filtering.

So the idea is that the computer automatically optimizes the mask to maximize the separation of the signal and the noise. This approach has not been widely practiced in TEM but clearly holds enormous promise. By varying the mask, this approach can prevent the analysis of a defect layer being dominated by the bulk information. Since the approach is quite straightforward signal processing, we will just illustrate an example found in the analysis of a simulated $\Sigma = 5$ grain boundary with an extra period along the boundary. To test the analysis, white noise was added to a calculated image to give the image shown in Figure 30.24A. The power spectrum (computer-generated DP) of the micrograph is shown in Figure 30.24B. The adaptive filter and the filtered image are shown in Figure 30.24C and D. The important feature of the adaptive filter is that it was created as such because the computer detected the doubling of the periodicity, which is *only* present in the grain boundary. Secondly, the mask consists of elongated openings which we know we need when analyzing the grain boundary because of the shape effect.

30.15. REAL OR RECIPROCAL SPACE?

In principle we could equally well compare two images in reciprocal space rather than real space. However, while the Fourier transforms can generally be carried out much faster, the real-space approach has several advantages:

- Fourier analysis separates local information into sine and cosine functions which are delocalized. When we reassemble the real-space image, higher parts of the frequency spectrum will be lost which will degrade the resolution.

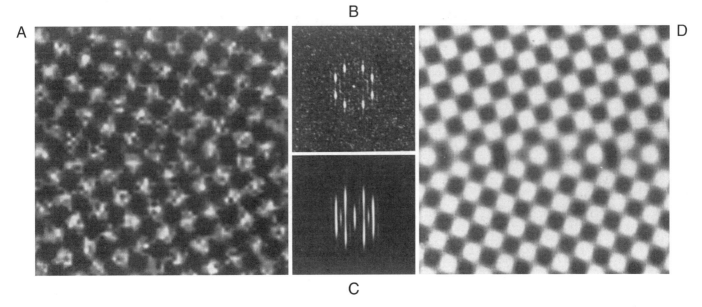

Figure 30.24. (A) White noise added to a calculated image of a $\Sigma = 5$ grain boundary; (B) The power spectrum of (A); (C) the adaptive filter; (D) the filtered image.

- We want to maintain information on the absolute value of the intensities.
- Real-space methods are visually more intuitive for most of us. We can easily see what we have removed in the process.
- The real-space approach allows us to choose any values of n and m in defining our templates. Fourier space prefers aspect ratios given by 2^n.

30.16. THE OPTICAL BENCH

Although not widely used now, the optical bench is still a useful instructional tool. A typical experimental set-up is shown in Figure 30.25. The laser provides a coherent source of illumination representing the electron beam. The negative acts as the specimen. If it contains a set of lattice fringes, these act as a diffraction grating and give rise to a row of spots on the screen placed at the back focal plane of the "objective" lens. The lens is thus performing an optical Fourier transform of the photograph. If you move the screen to the image plane, the fringes reappear. You can make different masks and place them at the back focal plane, or even create an "adaptive filter" by exposing a photographic film and using this as the template for your mask. These masks correspond to our objective apertures. Students will find it instructive to transform their instructor or another suitable photograph, examine the frequency spectrum, and investigate the resulting spatial effect of different masks. The detail in the image is quickly lost as you remove the high spatial frequencies. This corresponds to inserting a smaller aperture in the back focal plane of the objective lens, as illustrated in Figure 30.26. So Figure 30.26D is effectively a BF image: you lose a lot of information in such images!

Figure 30.25. A typical experimental set-up for an optical bench, with the mask in the back focal plane.

30 ■ QUANTIFYING AND PROCESSING HRTEM IMAGES

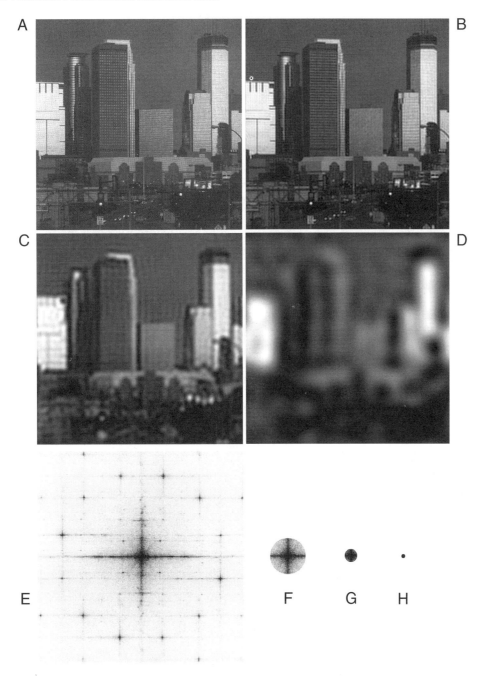

Figure 30.26. The effect of mask (aperture) size on a nonperiodic image of the Minneapolis skyline viewed from near the Guthrie Theater. (A–D) Reducing the aperture size, as indicated in the corresponding optical transforms (diffractograms) (E–H) reduces the image detail. The streaks in (E) arise from the edges of the photographs.

CHAPTER SUMMARY

We have been doing image processing for many years; it's called "dodging" in the photographic darkroom. You can even do this with a special enlarger. However, we have done very little quantitative imaging in materials science. The points you should remember when starting in the field are the following:

- Quantitative comparison of simulated and experimental images depends on both the simulation program and your experimental parameters.
- If you are going to use reciprocal-space techniques for quantitative analysis, you should let the computer design the optimum mask as part of this process; usually, it will not be a circular mask, especially if you are studying interfaces!
- The potential for image restoration is not limited by the signal mixing due to C_s and Δf. You can unscramble those effects. Ultimately, the limit is set by the signal-to-noise ratio in your image.
- You will notice the repeated use of the word "potential," where we don't mean $V(\mathbf{r})$! In many ways this chapter is a guide to the future of HRTEM and TEM in general. Some of the features won't be commonly available or optimized until the manufacturers realize their importance to the user.

There is always the possibility of removing information which is important. For example, Fresnel fringes often should be there! Beware of making reality match your simulation, rather than the reverse. In the same vein, we draw your attention to the conclusion of Hÿtch and Stobbs (1994), who found that they could only match their experimental and simulated images if they used a value for the specimen thickness which they knew was wrong! Their study emphasizes that, wherever possible, you should obtain independent measurements of the characteristics of your specimen and your machine. Remember the double-headed rhino in Figure 1.7; don't publish artifacts, even well-processed ones.

In this chapter we have discussed several different techniques used for processing TEM images. Several software packages are widely used by the TEM community and have been listed in Section 1.5. In its earliest application, image processing in TEM was almost exclusively applied to HRTEM images. This is no longer the case. Remember: always start with the best possible data. You can't always obtain a perfect image because your specimen might be beam-sensitive, or coated with oxide, and you need to be aware of these limitations when processing or quantifying the image. This chapter has given you a hint of what is possible and where the subject is developing. We recommend that you obtain the software and start experimenting.

REFERENCES

General References

Harrel, B. et al. (1995) *Using Photoshop for Macintosh,* Que, Indianapolis. A detailed description of the possibilities using Photoshop; this volume should also serve as a warning for all microscopists. Ask the question: has the image you are examining been processed, and if so, how?

Hawkes, P.W., Ed. (1980) *Computer Processing of Electron Microscope Images,* Springer-Verlag, New York. This collection of articles is for advanced students.

Hawkes, P.W., Ed. (1992) *Signal and Image Processing in Microscopy and Microanalysis,* Scanning Microscopy Supplement **6**, SEM, Inc. AMF, O'Hare, Illinois. This volume and its 1988 and 1996 companions are required reading for this subject.

Hawkes, P.W., Ottensmeyer, F.P., Saxton, W.O., and Rosenfeld, A. Eds. (1988) *Image and Signal Processing for Electron Microscopy,* Scanning Microscopy Supplement **2**, SEM, Inc. AMF, O'Hare, Illinois.

Hawkes, P.W., Saxton, W.O., and Frank, J., Eds. (1996) *Image Processing,* Scanning Microscopy Supplement, SEM, Inc. AMF, O'Hare, Illinois.

Russ, J.C. (1990) *Computer-Assisted Microscopy,* Plenum Press, New York. Chapter 3 is particularly relevant. Subsequent chapters give detailed analyses of the topics introduced here.

Russ, J.C. (1995) *The Image Processing Handbook,* 2nd edition, CRC Press, Boca Raton, Florida. A beautiful, comprehensive, and an essential component of any EM lab (or home).

Specific References

Dorset, D.L. (1995) *Structural Electron Crystallography,* Plenum Press, New York.

Downing, K.H. (1992) Scanning Microscopy Supplement **6**, ibid., p. 405.

Erickson, H.P. and Klug, A. (1971) *Phil. Trans. Roy. Soc. Lond.* **B261**, 105.

Frank, J. (1980) in *Computer Processing of Electron Microscope Images,* p. 187.

Hÿtch, M.J. and Stobbs, W.M. (1994) *Ultramicroscopy* **53**, 191.

King, W.E. and Campbell, G.H. (1993) *Ultramicroscopy* **51**, 128.

King, W.E. and Campbell, G.H. (1994) *Ultramicroscopy* **56**, 46.

Kirkland, E.J., Siegel, B.M., Uyeda, N., and Fujiyoshi, Y. (1982) *Ultramicroscopy* **9**, 65.

Kirkland, A.I., Saxton, W.O., Chau, K.-L., Tsuno, K., and Kawasaki, M. (1995) *Ultramicroscopy* **57**, 355.

Kisielowski, C., Schwander, P., Baumann, F.H., Seibt, M., Kim, Y., and Ourmazd, A. (1995) *Ultramicroscopy* **58**, 131.

Koster, A.J., van den Bos, A., and van der Mast, K.D. (1988) Scanning Microscopy Supplement **2**, p. 83.

Koster, A.J. and de Ruijter, W.J. (1992) *Ultramicroscopy* **40**, 89.

Krivanek, O.L. (1976) *Optik* **45**, 97.

Krivanek, O.L. and Mooney, P.E. (1993) *Ultramicroscopy* **49**, 95.

Möbus, G., Necker, G., and Rühle, M. (1993) *Ultramicroscopy* **49**, 46.

Moré, J.J. (1977) in *Lectures Notes in Mathematics* (Ed. G.A. Watson), p. 630, Springer, Berlin.

Moré, J.J., Garbow, B.S., and Hillstrom, K.E. (1980) User Guide for MINIPACK-1.

Ourmazd, A., Baumann, F.H., Bode, M., and Kim, Y. (1990) *Ultramicroscopy* **34**, 237.

Paciornik, S., Kilaas, R., Turner, J., and Dahmen, U. (1996) *Ultramicroscopy* **62**, 15.

Saxton, W.O. (1992) Scanning Microscopy Supplement **6**, p. 405.

Saxton, W.O. and Koch, T.L. (1982) *J. Microsc.* **127**, 69.

Saxton, W.O., Smith, D.J., and Erasmus, S.J.J. (1983) *J. Microsc.* **130**, 187.

Tang, D., Jansen, J., Zandbergen, H.W., and Schenk, H. (1995) *Acta Cryst.* **A51**, 188.

Trus, B.L., Unser, M., Pun, T., and Stevens, A.C. (1992) *Scanning Microscopy Supplement* **6**, p. 441.

van Heel, M., Winkler, H., Orlora, E., and Schatz, M. (1992) *Scanning Microscopy Supplement* **6**, p. 23.

Zou, X. (1995) *Electron Crystallography of Inorganic Structures*, Chemical Communications, Stockholm University, Stockholm.

Other Imaging Techniques

31

31.1. Stereo Microscopy	531
31.2. 2½D Microscopy	532
31.3. Magnetic Specimens	534
31.3.A. The Magnetic Correction	534
31.3.B. Lorentz Microscopy	535
31.4. Chemically Sensitive Images	538
31.5. Imaging with Diffusely Scattered Electrons	538
31.6. Surface Imaging	540
31.6.A. Reflection Electron Microscopy	540
31.6.B. Topographic Contrast	540
31.7. High-Order BF Imaging	541
31.8. Secondary-Electron Imaging	541
31.9. Backscattered-Electron Imaging	542
31.10. Charge-Collection Microscopy and Cathodoluminescence	543
31.11. Electron Holography	543
31.12. *In Situ* TEM: Dynamic Experiments	546
31.13. Other Variations Possible in a STEM	547

CHAPTER PREVIEW

What we've discussed in the preceding nine chapters comprises "classical" TEM imaging, based on BF or DF techniques. Diffraction contrast, phase contrast, and to a lesser extent mass-thickness contrast, are the mechanisms we use to characterize our specimens. We control the contrast by inserting the objective aperture, or a STEM detector, and excluding or collecting electrons that have been scattered by the different processes. However, there are variations to the standard ways in which we can extract more information from a TEM image and in this chapter we'll present a brief overview of some of them. Most of these operational modes that we'll discuss here are somewhat esoteric and have rather specialized applications. Nevertheless, you should know about them because they may be just what you need to solve your particular problem. There's no importance to the order in which we go through the various modes, but we'll cover modifications to conventional parallel-beam TEM imaging as well as those techniques that require STEM and use some of the electron detectors we discussed in Chapter 7. It turns out, however, that the various procedures are often feasible in either TEM or STEM mode.

This is a bit of a potpourri of a chapter but the techniques are not, to our knowledge, gathered together in any other text. The descriptions will, of necessity, be brief but we'll reference suitable source material so you can follow up if you really want to try the technique for yourself.

Other Imaging Techniques

31.1. STEREO MICROSCOPY

You should have realized by now that any TEM or STEM image is a two-dimensional projection of a 3D specimen and this is a fundamental limitation. Sometimes we can discern differences in diffraction-contrast images of certain defects, depending on whether the defect is intersecting the top or the bottom of the foil, but generally we lose the depth dimension. To regain this depth information we use stereo microscopy, but only for features showing mass-thickness or diffraction contrast. We cannot use stereo for phase-contrast imaging because the essential experimental step, tilting the specimen, changes the phase contrast and the projected potential of the specimen. So any stereo effect in the image is lost. You may need stereo microscopy if, for example, you want to know whether precipitates have formed on your specimen surface rather than in the interior, or if you want to see how dislocations are interacting with each other.

Stereo imaging works because your brain gauges depth by simultaneously interpreting signals from both your eyes, which view the same scene from slightly different angles (about 5°), giving a parallax shift. So in the TEM, if you take two pictures of the same area but tilted a few degrees relative to each other, then present the two images simultaneously to your brain using a stereo viewer, you'll see a single image in which the different depths of the features are apparent. In fact, some people are able to see the stereo effect without the aid of a viewer and some people are incapable of discerning the effect at all.

> You have to make sure that the field of view, the contrast, and the magnification in each of the two images are the same.

To see in stereo, the two images should be separated by ~60 mm, but in practice it's often sufficient just to move the pictures relative to each other until your eye and brain seize on the effect.

A couple of points are worth noting before we describe the method. First, if the features you want to observe show diffraction contrast, then the only way to maintain contrast is to tilt along a Kikuchi band, keeping both **g** and **s** fixed; so tilt while looking at the DP. This procedure almost invariably requires a double-tilt stage and may be difficult, or impossible, if your specimen is heavily deformed. If you just want to measure the foil thickness, any tilt is sufficient and contrast does not have to be maintained. Second, if you want to be pedantic, there is a right and a wrong way to view the stereo images. You have to present the images in the same way that your eyes would see a scene, that is, the two images have to be correctly positioned, otherwise the brain will interpret depth the wrong way round. If you are trying to perceive the true surface topography (such as with SE images) using SEM or STEM images, then the choice of which image goes into the left eye and which into the right is crucial. (See any SEM text, such as Goldstein *et al.* (1992), for more details on stereo viewing.) Of course, for TEM images this difference is irrelevant. TEM applications are reviewed by Hudson (1973) and a whole set of related papers appears in that same issue of *Journal of Microscopy*.

So if you want to take a stereo pair, follow these steps:

- Select the region of interest, making sure that the specimen is eucentric.
- Record an image (BF, DF, or WBDF, it doesn't matter, although usually the BF image is used).
- Tilt the specimen by at least 5° (much higher tilts give larger parallax shift, but it's more difficult to keep the focus and the diffraction-contrast constant).
- Ensure that the whole field of view didn't move while you tilted. If it did, translate it back to its original position (using the beam stop as a point

of reference if you wish). All features in the image should be shifted slightly relative to each other and it is this parallax shift which your brain interprets in stereo.

- If the area is now out of focus (it will be if you had to use the second, noneucentric, tilt axis), refocus *using the specimen-height (z) control;* otherwise you'll change your image magnification. Obviously, computerized stages will help in this respect.
- Take another photograph.
- Develop the images and observe them under a stereo viewer.

Figure 31.1 shows a pair of BF images showing precipitates. If the images are correctly spaced and you look through a stereo viewer, then you should be able to see the relative depth of the precipitates. You can purchase cheap cardboard stereo viewers through any EM supplier. Although a proper stereo viewer is an expensive optical tool, you can use it to calculate the relative depth (Δh) of a feature in a stereo pair, since

$$\Delta h = \frac{\Delta p}{2M \sin \frac{\phi}{2}} \quad [31.1]$$

where Δp is the parallax shift between the same feature in the two images tilted by ϕ at a magnification M. Be careful how you define ϕ because some microscopists define the tilt angle as $\pm \phi$, in which case $\sin \phi/2$ becomes $\sin \phi$. For true depth determination you need to deposit some recognizable feature on the surface, such as gold islands, but that is not usually important in TEM images and relative depth is often sufficient. For quantitative stereo measurements you have to enter the field of stereology, which is a discipline on its own and something that we can't cover at all, so you need to go and read an appropriate text, such as Russ (1990).

31.2. 2½D MICROSCOPY

This is one of the few examples of imaginative terminology in the TEM field, and is due to Bell (1976). If there are diffraction spots in the SAD pattern which are too close together to give separate DF images, center the objective aperture around them all. Then, if you view two DF images taken at different focus settings through a stereo viewer, you see features at different apparent depths. However, you're seeing a pseudo-stereo technique because the "depth" difference is due to a difference in **g**, not a true depth difference at all: hence the term "2½D" or "not quite

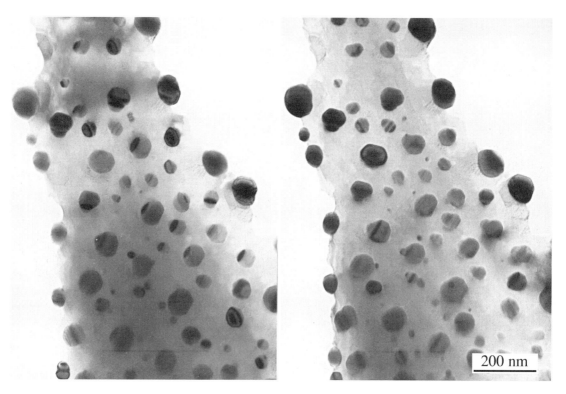

Figure 31.1. A stereo pair of spinel precipitates in a NiO-Cr_2O_3 specimen which shows the relative depth of the precipitates. You can see the small parallax shift between the two images.

31 ■ OTHER IMAGING TECHNIQUES

3D." The technique is also called by the rather more staid term "through-focus DF."

Bell developed a simple theory explaining how a change in focus Δf introduces a parallax shift in the image, y

$$y = M \Delta f \lambda \mathbf{g} \qquad [31.2]$$

$$\Delta y_{12} = \Delta f \lambda \mathbf{g}_{12} \qquad [31.3]$$

where M is the magnification, λ is the electron wavelength, and $\mathbf{g}$ is the diffraction vector. So, as you change focus, the parts of the image coming from different diffracted beams shift relative to each other by an amount, Δy.

The term $\mathbf{g}_{12}$ ($= \mathbf{g}_1 - \mathbf{g}_2$) is the vector between the spots arising from features 1 and 2. It is this parallax shift which introduces a stereo effect in the same manner as one introduced through tilting.

When would you need to use such a technique? Well, if you're dealing with a multiphase specimen in which many of the phases have similar structures and/or lattice parameters, then there would be many closely spaced diffraction spots. In these circumstances, if you succeed in separating the spots with a small objective aperture, the selected spot will undoubtedly be very close to the aperture and the image will display serious aberrations. It's more likely that you will be unable to select only one spot in the aperture.

You must start by setting up a CDF imaging configuration, but one in which several diffraction maxima are contained within the objective aperture. In the normal CDF image, therefore, all the parts of the specimen diffracting into the several spots will appear bright. However, if you take two images at different defocus conditions, the features diffracting into different spots will shift by slightly different amounts, given by equation 31.3. This relative shift between the two images is perceived by your brain as a relative depth difference when you view the images simultaneously through a stereo viewer. You might argue that your images are going to be out of focus and, strictly speaking, you're right, but remember that in the TEM there's a large depth of field (see Section 6.7) and it's easy to keep both images reasonably in focus over quite a range of objective lens excitation, so long as the magnification isn't too high. The greater the change in the objective lens current, the better your ability to resolve similarly diffracting features, so this process works better at lower magnifications where the depth of field is greater.

So the experimental steps are as follows:

- Select the area of interest making sure the specimen is eucentric.
- Tilt the beam so that the several diffraction spots cluster around the optic axis; center the objective aperture around this group of spots.
- Return to image mode and *underfocus* the objective lens until you see the clarity of the image begin to degrade; then overfocus one click back to maintain focus.
- Record an image.
- *Overfocus* the objective lens until the image begins to lose clarity again, then underfocus back one click.
- Record another image.
- Develop the images and view them in a stereo viewer.

Figure 31.2 shows a pair of CDF images taken under these conditions. The DP, as you can see, has many closely

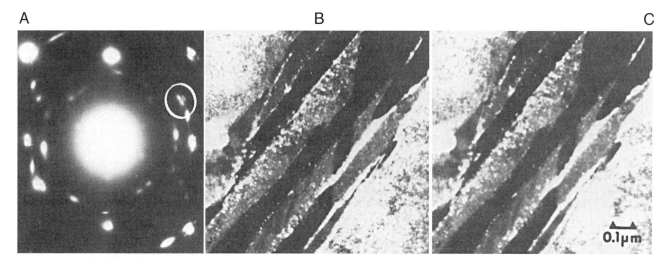

Figure 31.2. (A) SAD pattern of retained austenite and carbide precipitates in steel. (B,C) Stereo pair showing $2\frac{1}{2}$D effect in which the relative depths of the austenite and carbide are related to the position of their diffraction spots in the objective aperture in (A).

spaced spots. This example and many more are given in a well-illustrated review by Sinclair *et al.* (1981). If you look at the images through a stereo viewer, different bright regions will appear at different "heights." If you read the review, you'll find that it is quite easy to determine then which features in the image are responsible for which diffraction spot.

31.3. MAGNETIC SPECIMENS

If you happen to be looking at a magnetic material, then this can cause major problems for both you and the TEM and you might want to switch to studying aluminum. However, if you're patient and want a challenging task, then you can learn to correct for the magnetic disturbance introduced by the specimen. You can also get more information by making use of the interaction of the electron beam with the magnetic field of the specimen. TEM study of magnetic effects has seen somewhat of a resurgence since the discovery of high-T_c superconductors, and the growth of magnetic recording media.

First, we'll look at how to get the best images from your magnetic specimen and then we'll describe a couple of specialized imaging techniques that allow you to see either the domains or the domain walls, which are relevant to both ferromagnetic and ferroelectric materials.

31.3.A. The Magnetic Correction

If your specimen is magnetic, its magnetic field will deviate the electron beam as it passes through and then the electrons that you use to form images will not be on the optic axis. All your images will be severely aberrated and shift when you try to focus them. You can minimize these effects as we'll now describe.

The most important step is to make your whole specimen as thin and small as possible to reduce its total magnetic field strength.

> You should not use self-supporting magnetic disk specimens unless your material is brittle and likely to break in the microscope.

However, small flake specimens are more easily pulled out from between support grids when you put the specimen into the objective lens. So, make sure your foil is well clamped into the specimen holder and use oyster grids for thin flakes. Insert the holder into the lens with the objective lens strength as low as possible; look at the current running through the lens and minimize it.

You'll find it very difficult to set a magnetic specimen to the eucentric height because, as you tilt it through the 0° position, the specimen tends to try and rotate and either gets pulled out of the holder (if you're unlucky) or shifts slightly in height and position.

> If you lose your specimen it is paramount that you stop, switch off the microscope, and get help.

The column must be split and the lost specimen retrieved, otherwise you may introduce a fixed astigmatism into the microscope and, in the extreme, you may end up with your specimen welded to the polepiece, which gets to be quite expensive. It is better to incur the wrath of your technical support staff when the objective polepiece is still undamaged.

So when you have found a general area of interest, set up the eucentric height by tilting the specimen, taking care *not* to tilt through 0°; keep the tilt range to one side of zero. When the eucentricity is reasonable (and it rarely gets perfect for magnetic materials), focus and center a recognizable feature on the screen, remove the objective aperture, then carry out the following steps:

- Underfocus C2 and the objective lens by one coarse step. The feature should move off center. If you lose illumination, you haven't underfocused C2 enough.
- Return the feature to the screen center using the DF beam tilt potentiometers. This brings the imaging beam onto the optic axis.
- Refocus the image, recondense the beam, and if the feature shifts, recenter it with the stage traverses.
- Repeat the procedure for objective lens overfocus, until the feature stays centered as the objective lens is moved through focus.

If you then go to the DP, you'll see that the 000 beam should now be on axis and you can put the objective aperture back in again and check the astigmatism. However (and here's the catch), if you tilt or traverse the specimen, then you will have to re-do the whole correction since the magnetic field will change as you move the specimen relative to the beam. You'll rapidly get accomplished at this procedure if you practice for a while. Now, if you want to do CDF imaging at the same time as doing the magnetic correction, then you need an extra set of external DF controls, which may be an option or may be built into your microscope. Then you center the image feature with one set of potentiometers and tilt in the required $hk\ell$ maximum with the other set. Figure 31.3 shows the image of a mag-

31 ■ OTHER IMAGING TECHNIQUES

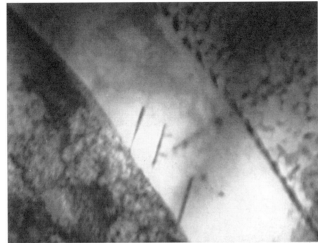

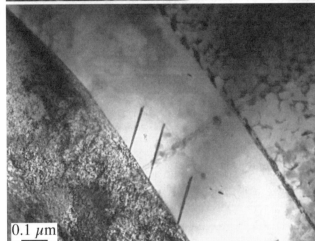

Figure 31.3. The effect of the magnetic correction on the quality of the image of an Fe-Ni meteorite: (A) before magnetic correction, (B) after correction.

netic specimen before and after the magnetic correction. It is more difficult to image magnetic specimens in STEM mode, because the scanning beam may interact differently with different areas of the specimen and so the image quality will be variable.

31.3.B. Lorentz Microscopy

When you've corrected for the magnetic field introduced by the specimen, you can image the magnetic domains if they're on the right scale to see in the microscope. This is a form of phase-contrast microscopy which we mentioned back in Chapter 27. The general term for this kind of imaging is called Lorentz microscopy (Chapman 1984) and it comes with two options.

Foucault Images: If there are several domains in the illuminated area, the electron beam will be deviated in different ways by different domains. This will result in a splitting of the diffraction spot, as shown in Figure 31.4A. It is useful to know immediately that the direction of magnetization is normal to the direction of spot splitting. You can then take a BF image using either all or one of the split 000 reflections. If you use all the spots, you can just see the domain walls (Figure 31.4B). But the domains bending electrons into the chosen spot will then appear bright and the other domains will appear darker, as in Figure 31.4C.

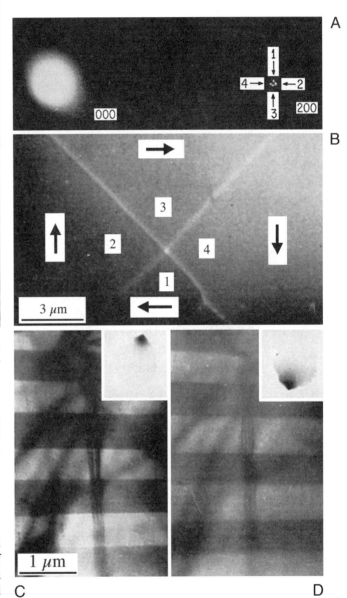

Figure 31.4. (A) Splitting of the 200 spot from Ni_3Mn due to the presence of magnetic domains. (B) Image taken from all of the split spots in (A) showing the four domains which scattered electrons into the various spots. (C,D) Foucault images of domains in Co formed by displacing the objective aperture to select one of two split spots as shown in the insets.

These images of the like domains are called Foucault images (perhaps after the French inventor of the pendulum used to demonstrate the earth's rotation) for reasons unknown to the authors. It is possible to see an analogous effect in STEM images if you use a detector that is cut into electrically isolated segments so different portions (usually halves or quadrants) pick up electrons coming through different domains. In STEM you can also add, subtract, or divide the various signals, as shown in Figure 31.5, because you are picking up each signal digitally (Craven and Colliex 1977). However, the intensity in Foucault images cannot be related quantitatively to the magnetic induction, so their only use is to give a rapid estimate of the domain size.

Fresnel Images: This option allows you to see the domain walls rather than the domains. As we discussed in Chapter 27, Fresnel imaging is named for another famous Frenchman whose micrographs were never in focus. If you over- or underfocus the objective lens, then the electrons coming through different domains will produce images in which the walls appear as bright or dark lines, which reverse contrast as you go through focus as in Figure 31.6. The contrast depends on whether the electrons going through the domains either side of the wall were deflected toward or away from each other, as shown schematically in the diagram (Figure 27.16A). From such images you can work out whether you're looking at Bloch walls, Néel walls, or cross-tie walls. Figure 31.7 explains why this technique is so important.

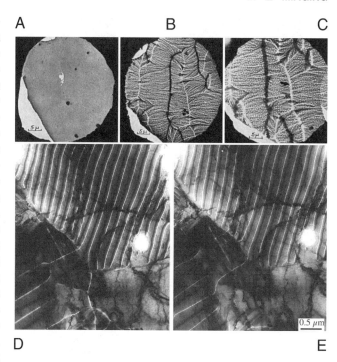

Figure 31.6. (A) In-focus image showing no magnetic contrast. (B,C) Fresnel-defocus images showing the magnetic domain walls which image as bright and dark lines and increase in width with defocus. (D) Underfocus and (E) overfocus images showing reversal of domain-wall contrast.

In an FEG TEM, the highly coherent source means that coherent Fresnel and Foucault (CF) imaging is possible. These techniques give quantitative measurements of the magnetic induction (Chapman *et al.* 1994).

If you want to do Lorentz microscopy, you have to decrease the field strength of your objective lens, which will otherwise dominate the internal field and thus control the domain size in the specimen. So, you either switch off the objective lens and use the intermediate lens for focusing, or use a specially designed low-field lens. TEM manufacturers offer appropriate objective lenses for Lorentz microscopy and it can also be done in STEM.

In addition to imaging the domains and the walls, you can also image the flux lines in the specimen if you evaporate Fe on the surface, just as you use Fe filings to delineate flux lines around a bar magnet. If you heat the specimen *in situ*, then the spot splitting decreases linearly to zero at the Curie temperature.

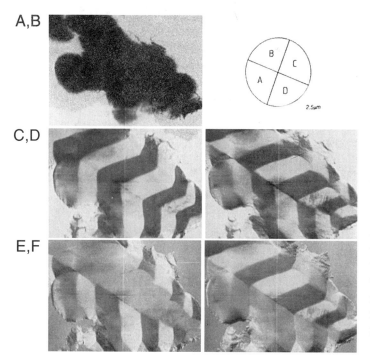

Figure 31.5. The use of a quadrant detector in STEM to differentiate regions of different magnetic induction in pure Fe. (A) The BF image from all four quadrants shown schematically in (B). (C–F) Images formed by different quadrant combinations. Regions showing the same intensity are regions of like induction.

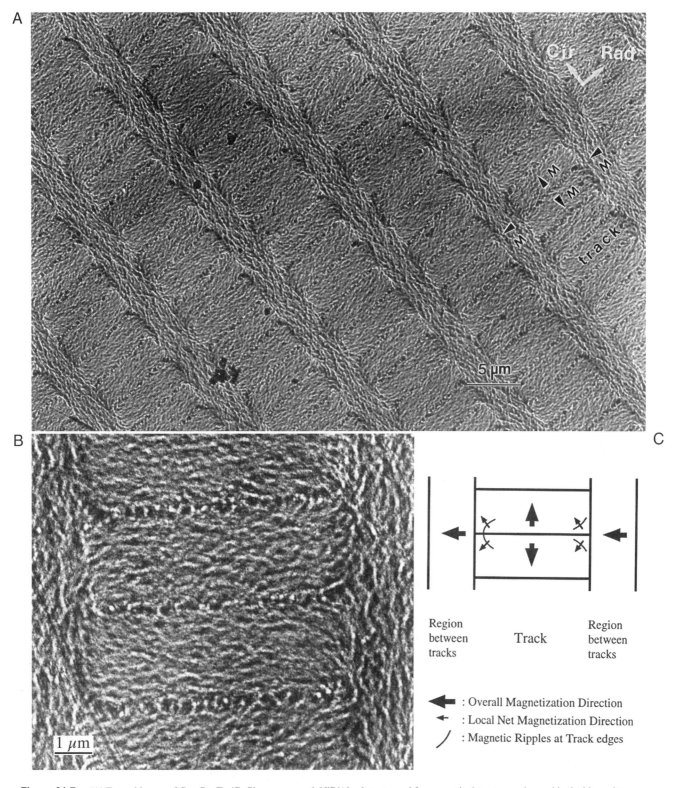

Figure 31.7. (A) Fresnel image of $Co_{84}Cr_{10}Ta_6$/Cr film on a smooth NiP/Al substrate used for magnetic data storage, imaged in the bits-written magnetic state. (B) Fresnel image at higher magnification. (C) Schematic of magnetic ripples at the track edges. The bits were in alternating direction of magnetization along the tracks in the circumferential direction of the hard disk, while the inter-track regions had remanent magnetization in the radial direction, perpendicular to that within the bits.

31.4. CHEMICALLY SENSITIVE IMAGES

In Section 16.4 we showed that in many materials the structure factor, F, for some reflections was sensitive to the difference between the atomic scattering amplitudes of the constituents. If we form a DF image using such a reflection, it will, in principle, be sensitive to changes in the composition of the material. The material systems which have been studied most extensively are related to GaAs and the other III-V compound semiconductors. There is great interest in partially replacing either the group III element or the group V element locally to produce superlattices and quantum wells, as illustrated in Figure 31.8. The contrast in the DF images will be brighter in thin specimens when the difference in the two atomic scattering amplitudes is large. Therefore, layers of $Al_xGa_{1-x}As$ appear brighter than the surrounding GaAs matrix.

This imaging technique can, in principle, be applied to many different materials; it is the same, in principle, as the use of superlattice reflections to image ordered regions (see Figures 16.5 and 16.6). We used the same information when discussing quantitative chemical lattice imaging in Section 30.11; we also use this effect when studying site location by ALCHEMI (see Section 35.8). In practice, it is often important to know how abruptly the composition changes, since this interface affects the properties of the material, so you may be interested in the change in contrast which occurs exactly at the interface. However, you should keep in mind our discussion (Section 23.11) of surface relaxations in the thin specimen which can influence any diffraction-contrast images.

31.5. IMAGING WITH DIFFUSELY SCATTERED ELECTRONS

This is just a variation on the theme of normal DF imaging. If your specimen contains noncrystalline regions which scatter electrons weakly compared to the diffraction spots, then the noncrystalline regions can be seen in strong contrast. To do this, you perform a CDF operation with the objective aperture centered away from any strong diffraction spots but at a position to intercept a fraction of the diffuse scatter. For example, in silicate glasses the diffuse scatter peaks radially at 3–4 nm^{-1} (Clarke 1979). As shown in Figure 31.9, a DF image reveals the amorphous regions with high contrast at the grain boundary in a ceramic bicrystal. However, you must be very careful when interpreting such images, as shown by Kouh *et al.* (1986).

Diffuse scatter can also arise from short-range ordering in the specimen due to either microdomains of ordered nuclei or local regions of increased order analogous to spinodal decomposition. These regions produce diffuse intensity maxima at positions that will eventually correspond to a superlattice spot when the short-range order has developed to long-range order (Cowley 1973a,b). A DF image from the diffuse scatter will reveal the short-range ordered regions as diffuse intensity maxima. Remember that we saw similar contrast effects, *but in the DPs* of short-range ordered materials in Figure 17.11.

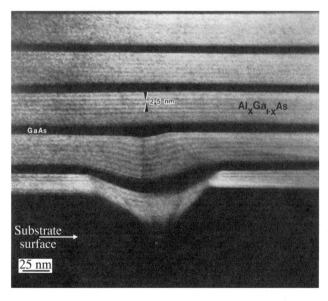

Figure 31.8. A chemically sensitive image of a GaAs-AlGaAs quantum-well structure. The composition of the AlGaAs is nonuniform because of growth fluctuations and the substrate surface was imperfectly covered by the first GaAs layer.

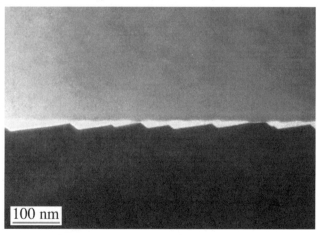

Figure 31.9. Diffuse-scatter DF image of a glassy material at a GB in a bi-crystal of Al_2O_3.

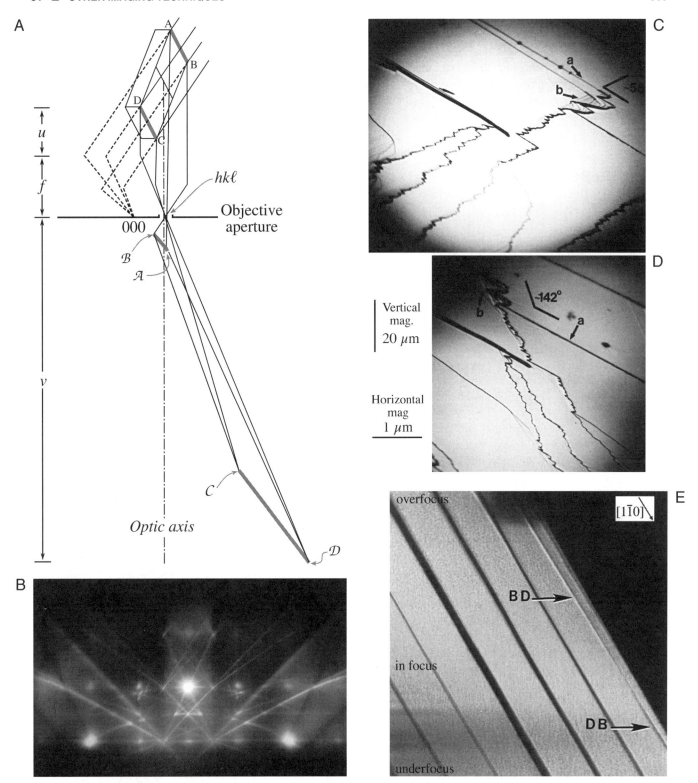

Figure 31.10. (A) Schematic diagram of the formation of an image using REM. (B) A 100 REM DP from a Si 001 surface. You can see the square array of spots but the 000 reflection has been blocked by the specimen itself. (C,D) REM images showing surface steps on a cleaved single crystal of GaAs. The two images are rotated with respect to one another and show the effect of the foreshortening. This emphasizes the care needed when interpreting foreshortened REM images. (E) REM image showing chemically sensitive contrast from quantum wells in GaAs/AlGaAs.

31.6. SURFACE IMAGING

We can get surface information in the TEM in a variety of ways. We can do reflection electron microscopy (REM) (Hsu 1992), or use a technique called "topographic contrast." We can also form an SE image in an SEM or STEM, as we describe later in the chapter.

31.6.A. Reflection Electron Microscopy

Reflection electron microscopy (REM) of surfaces requires that you mount your specimen in the holder so the beam hits at a glancing angle, as shown in Figure 31.10A. Since the electron is scattered from the surface, your specimen doesn't need to be thinned. The image is foreshortened by an amount which depends on the reflection used. Different parts of the specimen will be focused at different positions behind the lens. Once you have chosen an image plane, you can move the specimen in the z direction to focus different regions of the surface. A reflection high-energy electron diffraction (RHEED) pattern is generated by the surface layers of the specimen, as shown in Figure 31.10B. This diffraction geometry is exactly that which we used to derive Bragg's Law.

Once you've formed the DP, the experimental procedure is essentially the same as for conventional diffraction-contrast DF imaging. You insert an objective aperture to select a Bragg-reflected beam and form an REM image, as in Figures 31.10C and D. The images are strongly foreshortened in the beam direction but maintain the usual TEM image resolution in the plane normal to the beam which can make interpretation difficult; you'll see two different magnification markers in the two orthogonal directions. Note how different the two rotated images appear.

Ideally, you would like to have a very clean surface to simplify interpretation of the contrast. Of course, if the surface is not very flat, or is covered with a thick contamination or oxide layer, you won't learn much from the image. However, with care, you can use REM to study the surface of many different materials. All that you've learned about diffraction-contrast images will apply to REM images. For example, you can also detect chemically sensitive contrast as shown in Figure 31.10E. Here, the dark bands are layers of $Al_xGa_{1-x}As$ in a GaAs matrix; the contrast is sensitive to the actual value of x. Notice that this contrast is the reverse of what we saw in TEM using chemically sensitive reflections (De Cooman *et al.* 1984).

Since you can do REM using any regular TEM holder, you can easily heat or cool your bulk sample. In many ways, this is easier than using a thin transmission specimen since the sample is much more robust. *In situ* REM studies of Si provided a leap in our understanding of the reconstruction of Si surfaces. Yagi *et al.* (1987) showed the surface reconstruction taking place *in situ*. You'll find more details and examples in the books and journal issues listed in the general references.

31.6.B. Topographic Contrast

You can get a sense for the topography of your specimen by a neat technique which simply involves displacing the objective aperture until its shadow is visible across the region of the image that you're looking at. Around this area of the specimen you will see contrast which arises from an electron refraction effect (Joy *et al.* 1976) but can be simply interpreted in terms of thickness changes in the specimen. In Figure 31.11 you can easily see that the Fe_3O_4 particles are supported on top of the carbon film and you can also see that the carbon film is not flat. Although the displacement of the aperture introduces astigmatism into the image, it does not limit resolution at the relatively low magnification at which this image was taken. To carry out

Figure 31.11. Topographic contrast from Fe_3O_4 particles on a carbon film. Ripples in the film are clearly visible. The shadow of the displaced objective crosses the field of view.

an equivalent operation in STEM, you just displace the BF detector or shift the 000 diffraction disk so it falls half on and half off the detector.

31.7. HIGH-ORDER BF IMAGING

As we've taken some pains to point out, you get the strongest diffraction contrast in your images when **s** is small and positive. Under these circumstances it's easiest to see crystal defects and interpret their contrast. This strong dynamical contrast can mask the details of the defect and sometimes it is more important to lower the contrast. You can do this by operating under kinematical conditions. Rather than having one strong low-index spot in the DP, if you tilt to make a high-order spot bright then the overall contrast is less and the background is not dominated by bend contours or thickness fringes, and the effects are shown in Figure 31.12 (Bell and Thomas 1972). You can get a similar effect called "kinematical diffraction imaging" if you form a BF image when no $hk\ell$ spot is strong.

We get a similar effect in STEM imaging because, as we saw back in Section 22.3, dynamical-contrast effects are reduced when the reciprocity conditions between TEM and STEM are not fulfilled and all STEM images are more kinematical in nature.

Notice that Figure 31.12 was recorded at 650 kV. This tells you that high-order BF imaging can be a very useful technique when you're using an IVEM. You'll see there are subtle contrast effects that can occur at higher voltages. For example, in addition to improving the spatial localization of defect images, there are times when the 2**g**

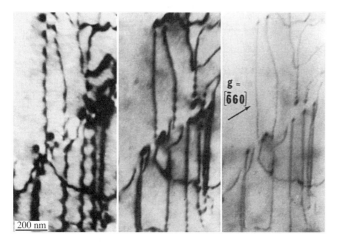

Figure 31.12. BF images of dislocations in GaAs taken with **g, 2g, 3g**. In the higher-order BF images the width of the defects decreases.

reflection gives better contrast than the usual **g** reflection. The point is, as usual, be prepared to experiment when you are sitting at the microscope. Many standard texts and many standard microscopists assume that you will be working at 100 kV, but only because that was all that was routinely available twenty years ago. The world is now different.

31.8. SECONDARY-ELECTRON IMAGING

SE images reveal the surface topography, which isn't much if you've polished your sample well, but is very important if you're looking at particulate specimens such as catalysts. You have to use a STEM if you're going to form SE images. If you look back to Figure 7.2, you'll see how the SE signal is detected by a scintillator-PM detector situated in the upper objective polepiece of a STEM. SEs generated in the top few nanometers of the specimen surface are confined by the strong magnetic field of the upper polepiece and spiral upward until they see the high voltage (~10 kV) on the aluminized surface of the scintillator. This design is different than the SEM, in which the SE detector is situated under the final polepiece, and in the STEM we get SE images of superior resolution and quality. There are several reasons for this:

- An SE image in an SEM invariably has noise contributions from BSEs which can enter the scintillator directly. But in the STEM there is no line of sight to the SE detector for the BSE, and so the SE signal in STEM lacks the BSE noise that exists in an SEM image.
- The brightness of a thermionic source in a STEM is higher than in an SEM, because of the higher kV, and so the SE signal will be correspondingly stronger.
- The C_s of STEM objective lenses is usually a lot smaller than for conventional SEMs. Therefore, SE images in a STEM are invariably of better quality than in an SEM because these two factors increase the *S/N* ratio.

The fourth advantage depends on knowledge of the different types of SE which we'll now discuss.

The presence of remote SE signals in an SEM also decreases the *S/N* ratio. As shown in Figure 31.13, the SE detector in an SEM can pick up four different types of secondary electrons, labeled SE I - SE IV (Peters 1984):

- The SE I signal is the only signal we want, since it emanates from the region around the

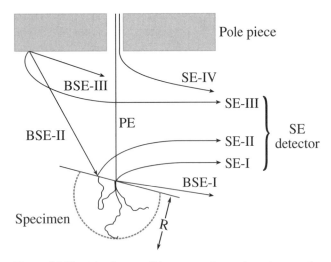

Figure 31.13. The four possible sources of secondary electrons that enter the detector in a conventional SEM. The SE I signal is the only desirable one since it comes from the probe region, but SE II from BSE electrons, SE III from the stage of the microscope and SE IV from the final aperture all combine to reduce the *S/N* ratio in an SE image.

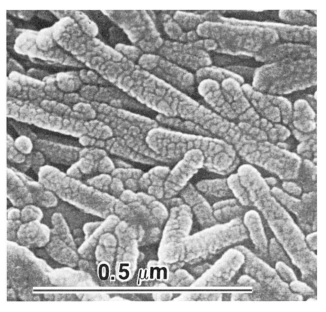

Figure 31.14. High-resolution SE image of coated magnetic tape in a TEM/STEM with an LaB_6 source at 100 kV showing ~2 nm spatial resolution.

probe and contains high resolution topographic information from that region only.

- The SE II signal comprises SEs generated by BSEs that emerge from the specimen some distance away from the beam. The only way to reduce SE II is to reduce the BSE fraction and we can do this by using thin specimens, as in STEM.
- The SE III component arises from parts of the microscope stage that are struck by BSE from the specimen. So, reducing the BSE fraction by using thin specimens also reduces SE III.
- The SE IV signal comes from SEs generated at the edges of the final probe-forming aperture which, in an SEM, often sits in the final lens. In a STEM, however, the C2 aperture performs this function and it is well away from the stage and the SE detector.

So in a STEM with a thin specimen, the ratio of desirable SE I to undesirable SE II - IV is higher than in a conventional SEM. If you look at a bulk sample in the STEM, then the SE II signal will be the same as in an SEM since the BSE yield does not vary with beam energy. Furthermore, the SE III signal may be slightly worse than in an SEM because of the smaller STEM stage. Experimentally, the net result of all these differences is that the STEM still provides higher-resolution SE images of a bulk sample than a conventional SEM, as shown in Figure 31.14. The highest-resolution FEG SEMs incorporate many of the design aspects of the STEM stage and can now produce SE image resolutions < 1 nm at 30 kV. A thermionic source STEM cannot better that performance, however an FEG STEM at 100 kV or higher should, in theory, offer the best possible SE images (Imeson 1987) and may even show atomic-level topographic information.

Despite these obvious advantages and the ready availability of STEMs with SE detectors, very few high-resolution studies of the surface topography of specimens have been carried out with a STEM. It appears that one of the major uses of the SE image is simply to find the hole in the TEM disk! Why this is so remains unclear, but we strongly encourage you to use SE images if they are available in your STEM. Of course, your specimen surface must be prepared carefully and kept clean, otherwise you'll image contamination, oxide, or some other artifact of your preparation process. If you have to coat your specimen because it's an insulator, then you should use a modern high-resolution, high vacuum coater to generate a continuous thin film of a refractory metal such as Cr. Don't use the more conventional Au-Pd coatings, which merely mask the fine detail of the specimen surface.

31.9. BACKSCATTERED-ELECTRON IMAGING

Remember that the BSE detector in a STEM is situated directly under the upper objective polepiece. In an SEM, the detector is in the same position and we get equal collection efficiency. Higher kV gives a brighter source, but specimen thickness dominates the yield and it is abysmal in thin specimens (0.4% in 100 nm of Au). So there is a very low total signal and the *S/N* is poor. Nevertheless, in a thin specimen we can still form BSE images in the STEM that show resolutions approaching the probe size, but only if a high-contrast specimen is used, e.g., Au islands on a C film, as shown in Figure 31.15. So in general, the BSE signal in a STEM offers no obvious advantages over a modern SEM with a high-efficiency scintillator or semiconductor detector. In fact, the manufacturer of dedicated STEMs does not even offer a BSE detector as an option, and it is doubtful if it is worth installing one on a TEM/STEM.

For both SE and BSE imaging in STEM, you can look at both electron transparent or bulk samples of the sort normally examined in a conventional SEM. A major drawback to the STEM compared with the SEM is the relative volume of the microscope stage. In an SEM stage you can insert samples up to several centimeters in diameter and several centimeters thick, which means that sample preparation from the original object is often minimal. The confined-stage region of a STEM means that, even with a specially designed holder for bulk samples, the largest specimen will be about 10 mm × 5 mm and not more than a couple of millimeters deep.

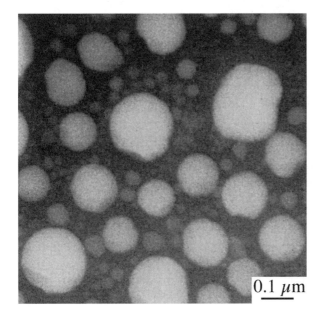

Figure 31.15. High-resolution STEM BSE image of Au islands on a carbon support film obtained in a TEM/STEM at 100 kV with an LaB_6 source. Resolution of ~7 nm is obtained, although the image quality is not very good because of the poor signal level.

31.10. CHARGE-COLLECTION MICROSCOPY AND CATHODOLUMINESCENCE

Charge-collection microscopy (CCM), otherwise known as electron beam induced conductivity (EBIC), and the related phenomenon of cathodoluminescence (CL) are common techniques for the characterization of semiconductors in the SEM (Newbury *et al.* 1986). It is possible to do the same things in a STEM, but few have been brave enough to try. You learned back in Section 7.1 about semiconductor electron detectors in which the incident beam generates electron–hole pairs which are swept apart by the internal field of the p–n junction and not allowed to recombine. So if your *specimen* is a semiconductor, electron–hole pairs will be formed during the normal imaging process. You have to separate out the pairs by applying an external voltage through ohmic contacts evaporated on the surface of your foil, then you can use the electron charge pulse to generate a signal on the STEM screen. The signal is strong wherever the pairs are separated, and weak at recombination centers such as dislocations and stacking faults. You can also measure the minority carrier diffusion length. Now in a STEM you can easily see the recombination centers by standard imaging techniques, so CCM is not really any great advantage, but it is essential in an SEM because the defects are subsurface.

If you don't separate the electron–hole pairs, they recombine and give off visible light. This light is an extremely weak signal, but it can be detected and dispersed by mirrors and spectrometers shown back in Figure 7.6. Again, recombination centers appear dark in CL images because most of the recombination is deep-level and nonradiative. The advantage of CL over CCM is that you don't have to coat your specimen to produce ohmic contacts, and you get a spectrum of light which contains information about doping levels and band-gap changes. However, you have to dedicate your STEM primarily to this imaging mode and it is a difficult and tedious technique. Early work in this area was pioneered by Petroff *et al.* (1978) and more recently by Batstone (1989). It is not obvious, however, that doing the work in transmission offers major advantages over studies of bulk samples in the SEM.

31.11. ELECTRON HOLOGRAPHY

Although the technique of electron holography achieved prominence in the early 1990s, Gabor had originally proposed the technique in 1948 as a way to improve the resolution of the TEM. The delay in its wide implementation was due to the lack of affordable, reliable FEG TEMs. An FEG is required so that the source will be sufficiently coherent; the FEG is then the electron equivalent of the laser. The topic is broader than you might have guessed, unless you've read the article by Cowley (1992) entitled "Twenty forms of electron holography." We'll discuss the general principles of the technique and refer you to the articles given in the general references for the details and more ideas.

The key feature is that, unlike conventional TEM imaging techniques, both the amplitude and the phase of the beam can be recorded. We can use this feature in two ways:

- The effects of C_s can be partially corrected; thus we can improve the resolution of the TEM.
- We can examine other phase-dependent phenomena, such as those associated with magnetism.

Several different forms of holography are possible (see Cowley 1992, for more).

- In-line holography.
- Single-sideband holography.
- Off-axis holography.

The approach which is mainly used in the TEM is the off-axis variation, as we'll discuss later. Figure 31.16 shows a hologram of a wedge-shaped crystal of Si oriented close to the [110] pole. The boxed region of the figure shows a set of fringes which are only 0.7 Å apart (Lichte 1992). These are not lattice fringes but do contain phase information relating to the diffracted beams. In order to analyze such images, you'll need the appropriate software although much of the original research used optical processing techniques; this is such a specialized branch of image processing and image reconstruction that we did not cover it in Chapter 30.

The principle of the technique is shown by the schematic in Figure 31.17. Here, $\chi(\mathbf{u})$ is the function we used in Chapter 29 to describe the effect of the objective lens aberrations and defocus. In conventional TEM, we choose conditions so that the specimen acts as a pure phase object and the imaginary part of $e^{i\chi(\mathbf{u})}$, i.e., the sine term, converts this phase information into an amplitude, which we record as the image. With holography, we can use the real part of the exponential too. A nice way to think of the process is that, for a real specimen, $\chi(\mathbf{u})$ mixes the amplitude, $\mathcal{A}$, and phase, ω, from the specimen to give amplitude, Λ, and phase, Ω, in the image. In conventional imaging, we record A^2 and lose all the information on Ω. With holography, we don't lose any information but we have to work hard to recover it. The original proposal by Gabor, to use holography

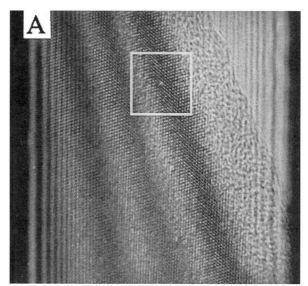

Figure 31.16. (A) Hologram of a [110]-oriented Si specimen; the broad bands are thickness fringes where the specimen is thickest at the lower left corner. (B) An enlargement of the boxed region in (A) showing with a 0.07 nm spacing.

to improve the resolution of the TEM, is being actively pursued wherever there is an FEG TEM (e.g., Harscher *et al.* 1995). Although holography has still not reached its full potential in this field, you can see the future by looking above at Figure 31.16.

In practice, electron holography is carried out using an FEG TEM which has been fitted with a beam splitter. The beam splitter is made by coating a thin glass fiber with metal to prevent it charging, and assembled to give the biprism, i.e., it's the biprism we discussed in Section 27.7 and can be < 0.5 μm in diameter. Part of the beam passes through the specimen while the other part forms the reference beam, as shown in Figure 31.18. You must be able to rotate either the biprism or the specimen. For holography,

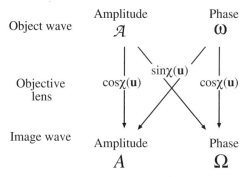

Figure 31.17. A schematic of how the objective lens mixes the amplitude and phase components of the specimen to form the amplitude and phase components of the image.

it's placed below the objective lens, e.g., at the position usually occupied by the SAD aperture rod.

> The essential feature of the experiment is that a reference beam passes outside your specimen and is then deflected by the biprism so that it interferes with the beam that passed through the specimen.

All that you then have to do is interpret the interference pattern, i.e., the electron hologram, but as you'd guess after reading Chapter 30, this process is not trivial. However, the technique does provide the possibility for "coherent processing" of the electron wave (Lichte 1992). In principle, all the required data can be recovered from one hologram

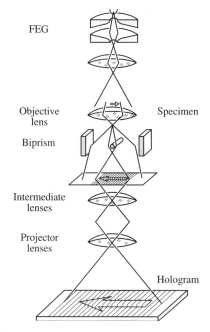

Figure 31.18. The operation of the biprism in electron holography.

but the technique is unlikely to replace conventional HRTEM in most applications.

We'll conclude this section by illustrating one of the unique applications of the technique, namely, the imaging of magnetization characteristics and flux lines which has been developed by Tonomura (1987). Figure 31.19 shows a series of images illustrating how magnetization

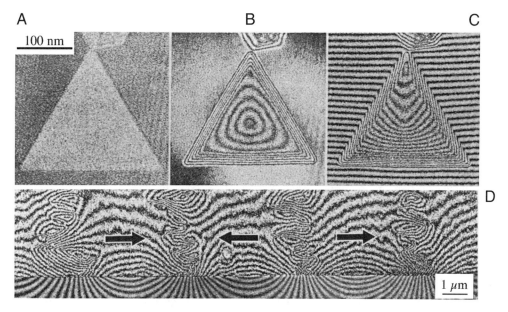

Figure 31.19. Holographic images of a magnetic Co particle showing (A) the reconstructed image, (B) the magnetic lines of force, and (C) the interferogram. (D) Lines of force in a magnetic recording medium.

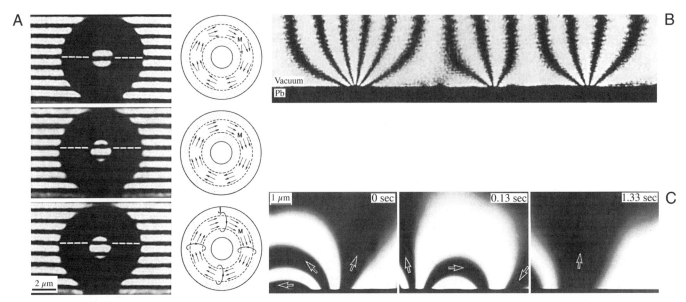

Figure 31.20. (A) Gradually cooling a ring (torus) of a superconducting material from 300K to 15K to 5K to demonstrate the quantization of flux. (B) Magnetic flux lines penetrating a superconducting Pb film. (C) Interference of fluxons trapped in superconducting Pb as time increases from 0 s to 0.13 s to 1.3 s.

patterns can be imaged both for single particles and for an actual recording medium. Tonomura's classic studies of the quantization of flux lines in superconducting materials is illustrated in Figure 31.20A. As the ring is gradually cooled, the phase shift due to the magnetization changes until it becomes superconducting, when the phase shift inside the ring becomes exactly π. The other images in Figures 31.20B,C show flux lines constricting as they enter the superconductor, but notice that you can clearly see these flux lines outside the specimen: the magnetic field outside the specimen also influences the electron beam.

31.12. *IN SITU* TEM: DYNAMIC EXPERIMENTS

We've mentioned several times that it is possible to do *in situ* experiments inside the TEM. In Chapter 8 we described holders you can use to heat the specimen, e.g., causing phase transformations, or to strain it to change the defect structure. In Chapter 30 we talked about CCD video cameras, which are the best way to record dynamic changes in the microstructure, and in this chapter we've discussed imaging moving flux lines in magnetic materials. *In situ* experiments remove the doubts that exist whenever you observe materials after heat treatment or after deformation and then try to infer what actually happened at temperature or during deformation. It is implicit in most

TEM investigations that cooling a heat-treated specimen to room temperature or removing the applied stress does not change the microstructure. Having made this assumption, we draw conclusions about what happened during our experiment. However, this assumption is clearly not valid for many situations. Nevertheless, we generally view our specimen at ambient temperatures and not under load.

In fact, there are good reasons why we rarely do *in situ* experiments. The main reason is that such studies are difficult to perform on thin specimens. As we indicated in Section 25.5, when the surface properties dominate the bulk, as is often the case in thin specimens, TEM images and analyses can be misleading. Surface diffusion is much more rapid than bulk diffusion and defects are subject to different stress states. The best way to overcome this limitation is to use much thicker specimens and this requires higher voltages. *In situ* experimentation was widely used in the 1960s and 70s when 1–3 MV instruments were first constructed and used to look through foils which were > 1 µm thick (see the text by Butler and Hale 1981). So *in situ* experiments were expensive!

However, the advent of 300–400 kV IVEMs and the construction of new 1.25-MV instruments in Stuttgart and Tsukuba has brought a resurgence of interest in *in situ* experimentation (see Rühle *et al.* 1994). Developments in electron optics, stage design, and recording media since the 1960s mean that combined HRTEM and *in situ* experimentation is now possible at intermediate voltages, as demonstrated in the beautiful images of Sinclair and Konno

Figure 31.21. (A–D) Four images taken from a video showing the reaction front of a Ge/Ag/Ge trilayer while heating the specimen *in situ* at 250°C. The time interval between the images was 8 s. The Ge crystal grows from the upper left without moving the Ag lattice.

(1994). This combination is a powerful tool, permitting the observation of reactions at the atomic level, such as the motion of individual ledges at interfaces, as in Figure 31.21, although lower-resolution images are no less impressive.

However, you must bear in mind that the experiments you perform *in situ* are taking place under conditions that still do not approach the bulk conditions experienced by many engineering materials. In particular, the high-keV electron flux brings an uncertainty to all *in situ* experiments, particularly if you want to try and infer kinetic data from measurements taken of reactions. So while *in situ* experiments can give you powerful demonstrations of real-time changes in materials, be wary of direct interpretation. You have to ensure you are controlling all the variables, and you must cross-check kinetic data with calculations to verify that, e.g., diffusion fluxes are consistent with the known temperature and not influenced by surface or vacancy effects. Having said that, it is nonetheless very useful to have access to an IVEM for this kind of experiment.

31.13. OTHER VARIATIONS POSSIBLE IN A STEM

We can form images with the characteristic X-rays and the energy-loss electrons that we detect with the appropriate spectrometers. Rather than discuss these here, we'll save the topics for the chapters devoted to XEDS (Chapter 35) and EELS (Chapter 40).

In a STEM then, we can pick up, in theory, all of the signals generated in a thin specimen and shown way back in Figure 1.3. We can reproduce all the conventional TEM imaging methods and some unconventional ones too, such as Z contrast as well as most of the specialized techniques in this chapter. Remember that for all the STEM signals we use a detector of one form or another. We can make detectors of different shapes and sizes, as we already described for looking at directional scattering from magnetic specimens. Remember also that the detectors permit us to digitize the signal so we can process it, manipulate it, and present it for viewing in ways that are impossible with analog

TEM images. Most STEM systems come with relatively basic image processing, such as black level, gain, contrast, brightness, gamma, Y modulation, and signal addition and subtraction. The computer system for X-ray and EELS analysis often has a lot of image analysis software thrown in. So there's a lot you can do with the STEM images that really falls into the category of digital image processing which you may study in specialist texts such as Russ (1990), after you've read Chapter 30.

CHAPTER SUMMARY

We've described only some of the many ways that we can manipulate the electron beam to produce different images and different contrast phenomena, and it is more than likely that there are still unknown methods awaiting discovery. For example, the introduction of a continuously variable detector collection angle in STEM is equivalent to using a continuously variable iris-type objective aperture in TEM and offers new imaging possibilities. Similarly, there is much to be learned from mutually complementary techniques in SEM and TEM (Williams and Newbury 1984).

We haven't discussed the use of annular apertures, conical illumination, pre- and post-specimen beam scanning, or rocking. All of these are possible, so you should always be prepared to try new things in the microscope and see what the effects are. Many advances of the sort we have described only came about by accident, but the microscopist was wise enough to see the effect and try and understand it, rather than dismiss it as unimportant.

REFERENCES

General References

Hsu, T., Ed. (1992) *J. Electron Microsc. Tech.* **20**, part 4. A special issue devoted to REM.

Larsen, P.K and Dobson, P.J., Eds. (1987) *Reflection High-Energy Electron Diffraction and Reflection Electron Imaging of Surfaces*, Plenum Press, New York.

Electron holography: The following key articles will give you a sense of the possible applications:

Gabor, D. (1949) *Proc. Roy. Soc. London* **197A**, 454; (1951) *Proc. Roy. Soc. London* **64**, 450. These are parts 1 and 2 of a pair of articles entitled "Microscopy by Reconstructed Wave Fronts."

Hansen, K.-J. (1982) *Adv. in Electronics and Electron Physics* **59**, 1.

Tonomura, A. (1987) *Rev. Mod. Phys.* **59**, 639; (1992) *Adv. Phys.* **41**, 59.

Ru, Q., Endo, J., Tanji, T., and Tonamura, A. (1991) *Appl. Phys. Lett.* **59**, 2372.

Spence, J.C.H. (1992) *Optik* **92**, 52.

Franke, F.-J., Hermann, K.-H., and Lichte, H. (1988) in *Image and Signal Processing for Electronic Microscopy*, (Eds. P.W. Hawkes, F.P. Ottensmeyer, A. Rosenfeld, and W.O. Saxton), p. 59, Scanning Microscopy Supplement **2**, AMF, O'Hare, Illinois.

Völkl, E., Allard, L.F., Datye, A., and Frost, B. (1995) *Ultramicroscopy* **58**, 97.

Steeds, J.W., Vincent, R., Vine, W.J., Spellward, P., and Cherns, D. (1992) *Acta Microsc.* **1.2**, 1. So that you don't forget the value of using an FEG for diffraction.

Specific References

Batstone, J. (1989) *Microbeam Analysis-1989* (Ed. P.E. Russell), San Francisco Press, San Francisco, California.

Bell, W.L. (1976) *J. Appl. Phys.* **47**, 1676.

Bell, W.L. and Thomas, G. (1972) *Electron Microscopy and Structure of Materials* (Ed. G. Thomas), p. 23, University of California Press, Berkeley, California.

Butler, E.P. and Hale, K.F. (1981) in *Practical Methods in Electron Microscopy* **9**, (Ed. A.M. Glauert), North-Holland, New York.

Chapman, J.N. (1984) *J. Phys. D* **17**, 623.

Chapman, J.N., Johnston, A.B., Heyderman, L.J., McVitie, S., and Nicholson, W.A.P. (1994) *IEEE Trans. Magn.* **30**, 4479.

Clarke, D.R. (1979) *Ultramicroscopy* **4**, 33.

Cowley, J.M. (1973a) *Acta Cryst.* **A29**, 529.

Cowley, J.M. (1973b) *Acta Cryst.* **A29**, 537.

Cowley, J.M. (1992) *Ultramicroscopy* **41**, 335.

Craven, A.J. and Colliex, C. (1977) *J. Microsc. Spectrosc. Electr.* **2**, 511.

De Cooman, B.C., Kuesters, K.-H., and Carter, C.B. (1984) *Phil. Mag. A* **50**, 849.

Goldstein, J.I., Newbury, D.E., Echlin, P., Joy, D.C., Romig, A.D. Jr., Lyman, C.E., Fiori, C., and Lifshin, E. (1992) *Scanning Electron Microscopy and X-ray Microanalysis*, 2nd edition, Plenum Press, New York.

Harscher, A., Lang, G., and Lichte, H. (1995) *Ultramicroscopy* **58**, 79.

Hudson, B. (1973) *J. Microsc.* **98**, 396.

Imeson D. (1987) *J. Microsc.* **147**, 65.

Joy, D.C., Maher, D.M., and Cullis, A.G. (1976) *J. Microsc.* **108**, 185.

Kouh, Y.M., Carter, C.B., Morrissey, K.J., Angelini, P., and Bentley, J. (1986) *J. Mat. Sci.* **21**, 2689.

Lichte, H. (1992) *Signal and Image Processing in Microscopy and Microanalysis,* Scanning Microscopy Supplement, **6**, p. 433, SEM Inc., AMF O'Hare, Illinois.

Newbury D.E., Joy. D.C., Echlin, P., Fiori, C.E., and Goldstein, J.I. (1986) *Advanced Scanning Electron Microscopy and X-ray Microanalysis,* p. 45, Plenum Press, New York.

Peters, K.R. (1984) *Electron Beam Interactions with Solids for Microscopy, Microanalysis & Microlithography,* (Eds. D.F. Kyser, H. Niedrig, D.E. Newbury, and R. Shimizu), p. 363, SEM Inc., AMF, O'Hare, Illinois.

Petroff, P., Lang, D.V., Strudel, J.L., and Logan, R.A. (1978) *Scanning Electron Microscopy* **1** (Ed. O. Johari), p. 325, SEM Inc., AMF, O'Hare, Illinois.

Rühle, M., Phillipp, F., Seeger, A., and Heydenreich, J., Eds. (1994) *Ultramicroscopy* **56**, (1-3).

Russ, J. (1990) *Computer-Assisted Microscopy: The Analysis and Measurement of Images,* Plenum Press, New York.

Sinclair, R. and Konno, T.J. (1994) *Ultramicroscopy* **56**, 225.

Sinclair, R., Michal, G.M., and Yamashita, T. (1981) *Met. Trans.* **12A**, 1503.

Williams, D.B. and Newbury, D.E. (1984) in *Advances in Electronics and Electron Physics,* (Ed. P.W. Hawkes), Academic Press, New York.

Yagi, K. Ogawa, S., and Tanishiro, Y. (1987) *Reflection High-Energy Electron Diffraction and Reflection Electron Imaging of Surfaces,* (Eds. P.K. Larsen and P.J. Dobson), p. 285, Plenum Press, New York.